I0504005

Arduino Smart Vacuum Cleaning Robot, Ping Pong Game, Call and Message utilizing, Private Chat Room, mobile phone Controlled Digital Code Lock

Arduino Smart Vacuum Cleaning Robot, Ping Pong Game, Call and Message utilizing, Private Chat Room, mobile phone Controlled Digital Code Lock

CONTENTS

ACKNOWLEDGMENTS

The writer might want to recognize the diligent work of the article group in assembling this book. He might likewise want to recognize the diligent work of the Raspberry Pi Foundation and the Arduino bunch for assembling items and networks that help to make the Internet of Things increasingly open to the overall population. Yahoo for the democratization of innovation!

INTRODUCTION

The Internet of Things (IOT) is a perplexing idea comprised of numerous PCs and numerous correspondence ways. Some IOT gadgets are associated with the Internet and some are most certainly not. Some IOT gadgets structure swarms that convey among themselves. Some are intended for a solitary reason, while some are increasingly universally useful PCs. This book is intended to demonstrate to you the IOT from the back to front. By structure IOT gadgets, the per user will comprehend the essential ideas and will almost certainly develop utilizing the rudiments to make his or her very own IOT applications. These included ventures will tell the per user the best way to assemble their very own IOT ventures and to develop the models appeared. The significance of Computer Security in IOT gadgets is additionally talked about and different systems for protecting the IOT from unapproved clients or programmers. The most significant takeaway from this book is in structure the tasks yourself.

1. DIY LED VU METER AS ARDUINO SHIELD

VU Meter or Volume Meter is prevalent and fun task in Electronics. We can consider the volume meter as an equalizer which is available in the Music frameworks. In which we can see the moving of LEDs as indicated by the music, in the event that music is uproarious, at that point equalizer go to its pinnacle and more LEDs will sparkle, and in the event that music is low, at that point lesser number of LEDs will shine. Volume Meter (VU) is a marker or portrayal of the force of sound level over LEDs and can likewise fill in as a volume estimation gadget.

Beforehand we assembled the VU Meter without utilizing Microcontroller and sound information was taken from Condenser Mic. This time we are building VU Meter utilizing Arduino as well as taking the sound contribution from 3.5 mm jack, with the goal that you effectively give sound contribution from your Mobile or Laptop utilizing AUX link or 3.5 mm sound jack. You can undoubtedly fabricate it on Breadboard yet here we are structuring it on PCB as an Arduino Shield utilizing EasyEDA online PCB test system and originator.

Components Required:

- Arduino UNO

- Power Supply

- VU Meter Arduino Shield (Self Designed)

Segments for VU Meter Arduino shield:

- Burg strips

- 3.5mm Audio Jack

- LEDs

- SMD type Resistors 100 ohm (10)

Designing Volume Meter (VU) Shield for Arduino:

For structuring VU Meter Shield for Arduino, we have utilized EasyEDA, in which first we have planned a Schematic and afterward changed over that into the PCB design via Auto Routing highlight of EasyEDA.

EasyEDA is a free online apparatus and one stop answer for building up your hardware ventures effortlessly. You can draw circuits, mimic them and get their PCB design in only a single tick. It additionally offers Customized PCB administration, where you can arrange the structured PCB in exceptionally minimal effort. Check here the total instructional exercise on How to utilize Easy EDA for making Schematics, PCB designs, reproducing the Circuits and so on.

EasyEDA has as of late propelled its new form (3.10.x), in which they have presented numerous new highlights and improved the general client experience, which makes EasyEDA progressively simpler and usable for structuring circuits. New form incorporates: improved MAC experience, improved parts search exchange, update PCB format in a single tick, include configuration notes in an edge underneath schematic and some more, you can discover all the new highlights of EasyEDA variant 3.10 here. Further they are before long going to dispatch its Desktop variant, which can be downloaded and introduced on your PC for disconnected use.

We have made the Circuit and PCB structure of this VU Meter Shield open, so you can simply pursue the connection to get to the Circuit Diagram and PCB formats.

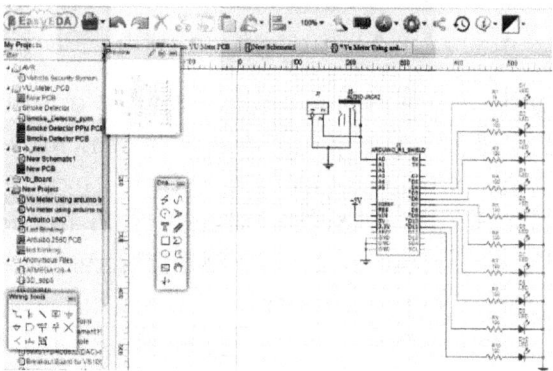

The following is the Snapshot of Top layer of PCB design from EasyEDA, you can see any Layer (Top, Bottom, Topsilk, bottomsilk and so on) of the PCB by choosing the layer structure the 'Layers' Window.

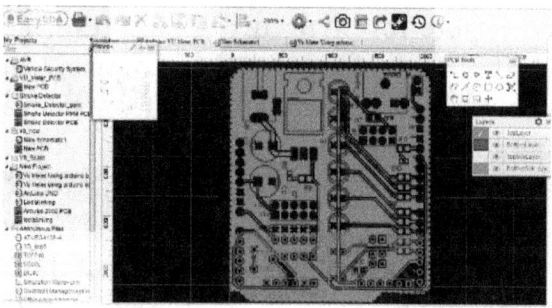

In case you discover any issue in utilizing EasyEDA, at that point look at our recently made 100 watt inverter circuit, where we have clarified the procedure bit by bit.

Ordering the PCB online:

In the wake of finishing the plan of PCB, you can tap the symbol of Fabrication yield, which will take you on the PCB request page. Here you can see your PCB in Gerber Viewer or download Gerber documents of your PCB and send them to any producer, it's likewise much simpler (and less expensive) to arrange it straightforwardly in EasyEDA. Here you can choose the quantity of PCBs you need to arrange, what number of copper layers you need, the PCB thickness, copper weight, and even the PCB shading. After you have chosen the entirety of the choices, click "Spare to Cart" and complete you request, at that point you will get your PCBs a couple of days after the fact.

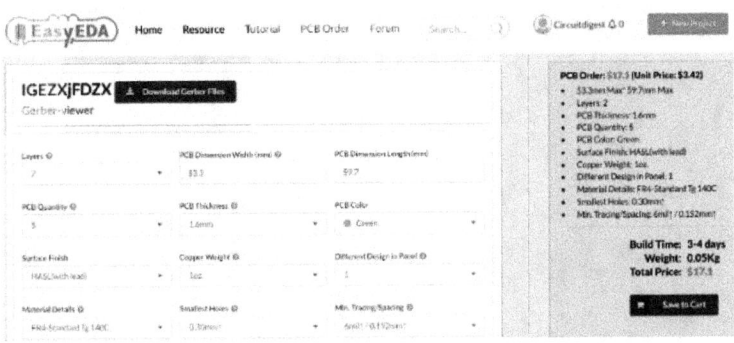

Following hardly any long periods of requesting the PCB, we got our VU Meter Arduino Shield PCB, and we found the PCBs in pleasant bundling and the nature of PCB is very great.

In the wake of getting the PCBs, we have mounted and bound all the necessary parts and burg strips over the PCB, you can have a last look here:

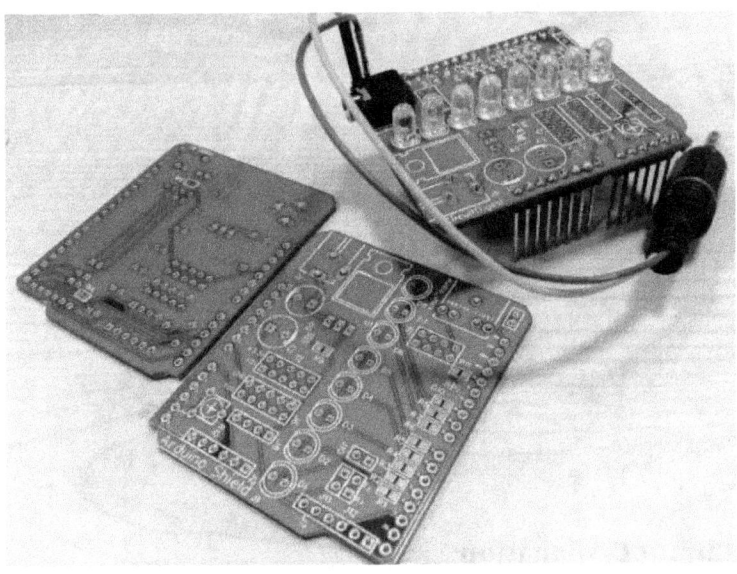

Presently we simply need to put this VU Meter Shield over the Arduino. Adjust the Pins of this Shield to the Arduino and solidly press it over the Arduino. Presently simply transfer the code to the Arduino and power on circuit and you are finished! Your VU Meter is prepared to move on music.

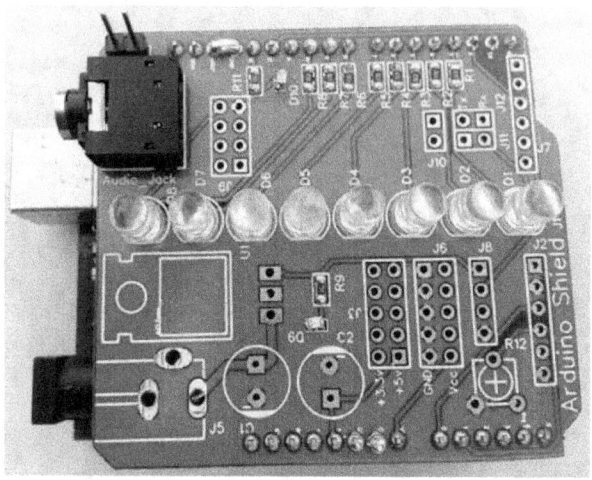

Circuit Explanation:

In this VU Meter Arduino Shield, we have utilized 8 LEDs, in which two LEDs are of Red shading for Higher Audio Signal, two Yellow LEDs are for intercede sound sign and four Green LEDs are for Lower sound Signal. We can include some more choice in this Shield by interfacing LCD, ESP8266 Wi-Fi module, DHT11 H&T Module, voltage controller, more VCC, +5v, +3.3v and GND pins. Be that as it may, here in show of this venture we have amassed just LEDs, sound jack and power LED. Here in this shield, we have utilized some SMD segments that are resistors and LEDs. Likewise we have 2 choices to apply sound sign to this board are immediate to pins or by utilizing sound jack.

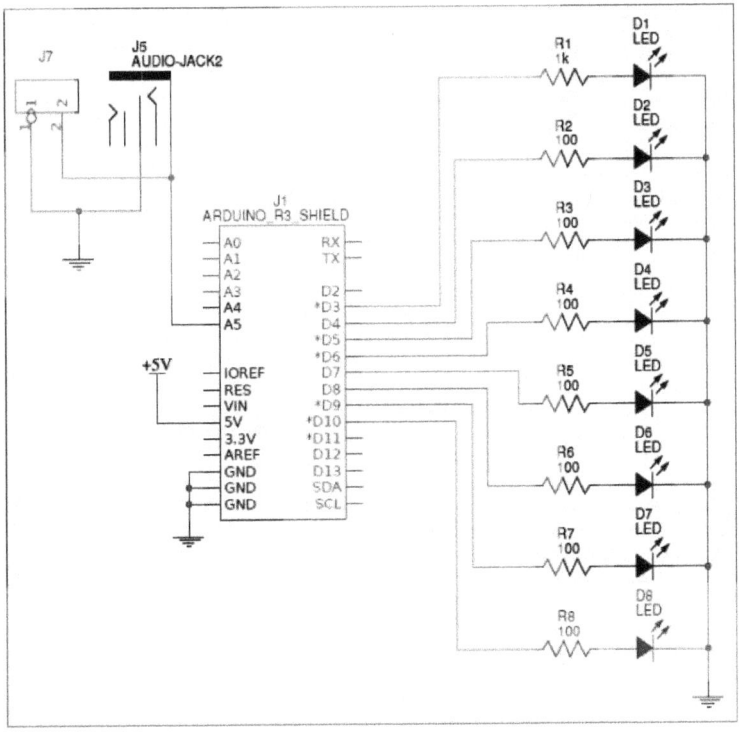

Circuit for this venture is extremely straightforward, we have an associated 8 LEDs at stick numbers D3-D10. Sound Jack is legitimately associated at simple stick A5 of Arduino.

On the off chance that you have to interface LCD, at that point you can associate the LCD at J1 and J7 (see underneath circuit) with associations like lcd(14, 15,16,17,18,2).

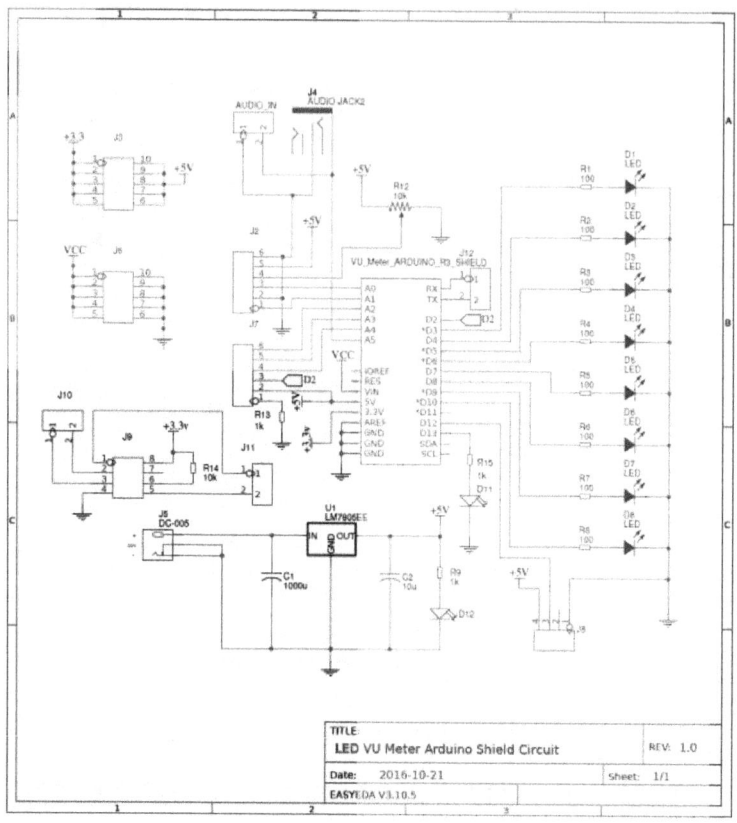

TITLE		
LED VU Meter Arduino Shield Circuit		REV: 1.0
Date: 2016-10-21		Sheet: 1/1
EASYEDA V3.10.5		

Programming Explanation:

Program of this Arduino VU Meter is exceptionally simple. Here in this code we haven't given any name to specific LED. I simply remember the association and compose code legitimately.

In the given void arrangement() work we instate the yield pins for LEDs. Here we can see a for circle wherein we instate the estimation of i=3 and run it to 10. Here i=3 is the third stick of Arduino and entire for circle will instate the stick D3-D10 of Arduino.

```
void setup()

{

  for(i=3;i<11;i++)

  pinMode(i, OUTPUT);

}
```

Presently in void circle() work we read the simple incentive from the A5 stick of Arduino and store that incentive in a variable to be specific 'esteem'. Presently this 'esteem' is isolated by 10 to get an outcome and this outcome is straightforwardly used to get stick no of Arduino utilizing for circle.

```
void loop()

{

  int value=analogRead(A5);

  value/=10;
```

```
for(i=3;i<=value;i++)

digitalWrite(i, HIGH);

for(i=value+1;i<=10;i++)

digitalWrite(i, LOW);

}
```

It tends to be clarified by model, as assume the simple worth is 50, presently isolate it by 10, we will get:

Worth = 50

Worth = esteem/10

Worth = 50/10 = 5

Presently we have utilized for circle like:

```
for(i=3;i<=value;i++)

digitalWrite(i, HIGH);
```

In above 'for' circle i=3 is D3 and Value=5 implies D5.

So it means circle will go from D3 to D5 and LEDs that are associated at D3, D4 and D5 will be 'ON'

What's more, in beneath 'for' circle i=value+1 implies value=5+1 implies D6 and i<=10 implies D10.

```
for(i=value+1;i<=10;i++)

digitalWrite(i, LOW);
```

Means circle will go from D6 to D10 and LEDs that are associated at D6-D10 will be 'OFF'.

So's the way we can manufacture our very own VU Meter Arduino Shield, in which LEDs will gleam as per the force of the sound. You can straightforwardly give contribution from your versatile or workstation by utilizing 3.5 mm sound jack or AUX link and mess around with the lovely lighting impact.

Code

```
int i=0;
int value1=0;
void setup()
{
 for(i=3;i<11;i++)
 pinMode(i, OUTPUT);
}
void loop()
{
 int value=analogRead(A5);
 value/=10;
 for(i=3;i<=value;i++)
 digitalWrite(i, HIGH);
 for(i=value+1;i<=10;i++)
```

```
 digitalWrite(i, LOW);
}
```

❖ ❖ ❖

2. 0-24V 3A VARIABLE POWER SUPPLY UTILIZING LM338

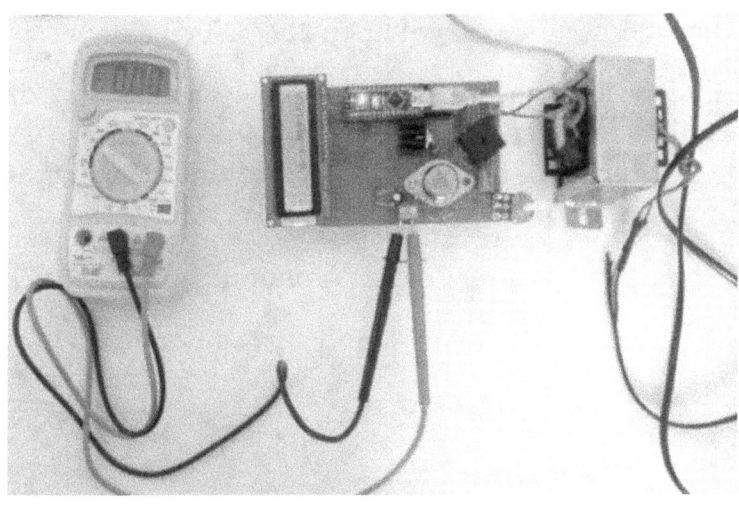

Batteries are commonly used to catalyst the Electronic Circuit and Projects, as they are effectively accessible and can be associated effectively. In any case, they depleted off rapidly and afterward we need new batteries, additionally these batteries can't give high current to drive an incredible engine. So to tackle these issues, today we are planning our own Variable Power Supply which will give Regulated DC voltage running from 0 to 24v with a most extreme current up to 3 Amps.

For the enormous majority of our Sensors and Motors we use voltage levels like 3.3V, 5V or 12V. Be that as it may, while the sensors require current in milliamps, engines like servo engines or PMDC engines, which run on 12V or more, require a high current. So we are working here the Regulated Power Supply of 3A current with the Variable voltage between 0 to 24v. Anyway in functional we got up to 22.2v of yield.

Here the voltage level is controlled with assistance of a Potentiometer as well as voltage worth is shown on LCD which will be driven by an Arduino Nano. Likewise look at our past Power supply circuits:

- 12v Battery Charger Circuit utilizing LM317?

- Variable Power Supply By Arduino Uno

- PDA Charger Circuit

Materials Required:

- Transformer - 24V 3A
- Dot board
- LM338K High Current Voltage Regulator
- Diode Bridge 10A
- Arduino Nano
- LCD 16*2
- Resistor 1k and 220 ohms
- Capacitor 0.1uF and 0.001uF
- 5K variable Pot (Radio Pot)

- 7812 Voltage Regulator
- Terminal Block
- Berg stick (Female)

How it works:

A RPS is one which changes over your AC mains into DC as well as directs it to our necessary voltage level. Our RPS utilizes a 24V 3A venture down transformer which is amended into DC utilizing a diode connect. This DC voltage is directed to our necessary level by utilizing LM338K and constrained by utilizing a Potentiometer. The Arduino and LCD are controlled by a low current rating Voltage controller IC like 7812. I will clarify the circuit bit by bit as we experience our task.

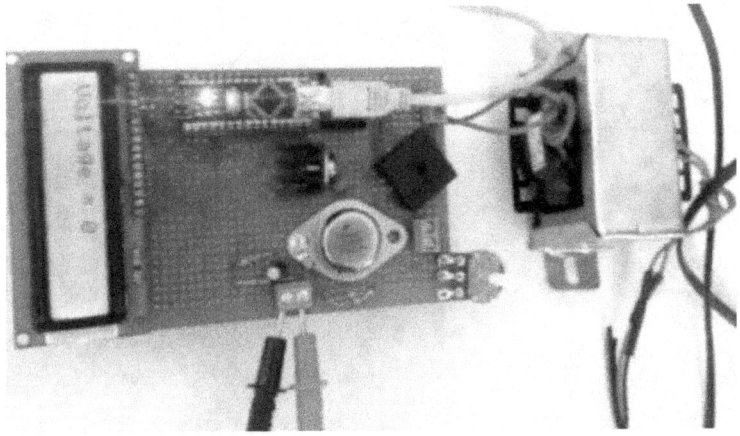

Connecting LCD with Arduino to Display Voltage Level:

How about we start with the LCD show. In the event that you know about LCD interfacing with Arduino, you can skirt this part and straightforwardly bounce to next area and on the off chance that you are new to Arduino and LCD, it won't be an issue as I will direct you with codes and associations. Arduino is an ATMEL controlled microcontroller pack which will help you in building ventures effectively. There are heaps of variations accessible however we are utilizing Arduino Nano since it is minimal and simple to use on a dab board

Numerous individuals have confronted issues in interfacing a LCD with Arduino, that is the reason we attempt this first so it doesn't destroy our venture in a minute ago. I have utilized the accompanying to begin with:

This Dot board will be utilized for our whole hardware; it is prescribed to utilize a female berg stick to fix the Arduino Nano with the goal that it could

be reused later. You can likewise confirm the working utilizing a breadboard (Recommended for apprentices) before we continue with our Dot board. There is a decent control by adafruit for lcd you can check it .The schematics for arduino as well as LCD is given beneath. Arduino UNO is utilized here for schematics, however not to stress the Arduino NANO and UNO have the equivalent pinouts and work the equivalent.

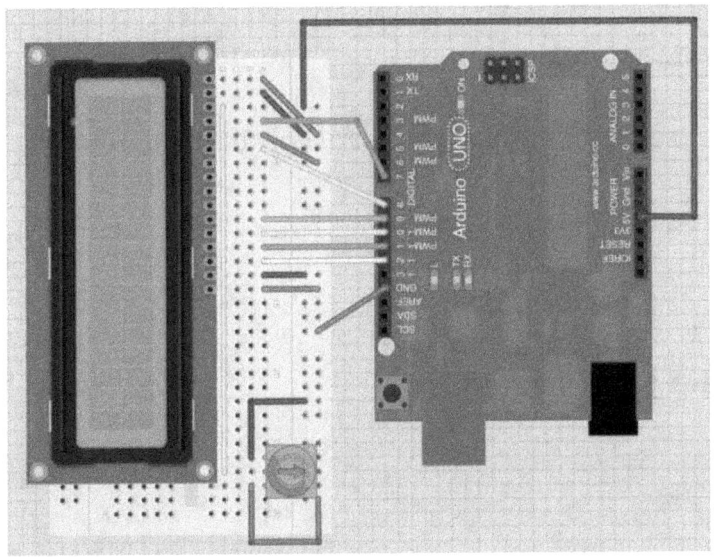

When the association the done you can transfer beneath code legitimately to check the LCD working. The header document for LCD is given by Arduino as a matter of course, don't utilize any unequivocal headers as they will in general give mistakes.

```
#include <LiquidCrystal.h>

// initialize the library with the numbers of the
interface pins

LiquidCrystal lcd(7, 8, 9, 10, 11, 12);

int a = 5;

void setup() {

  // set up the LCD's number of columns and rows:

  lcd.begin(16, 2);

  // Print a message to the LCD.

  lcd.print("hello, world!");

}

void loop() {

  // set the cursor to column 0, line 1

  // (note: line 1 is the second row, since counting
  begins with 0):
```

```
lcd.setCursor(0, 1);

// print the number of seconds since reset:

lcd.print(a);

}
```

This ought to get your LCD to work, yet in the event that regardless you face issues attempt the accompanying:

1. Check you sticks definition in the program.

2. Legitimately ground the third stick (VEE) and fifth stick (RW) of your LCD.

3. Ensure you LCD pins are put in the correct request, some LCD's have their pins is another bearing.

When the program works it should look something like this. On the off chance that you have any issues told us by remarks. I have utilized the smaller than usual USB link to control the Arduino for the time being, yet later we resolve it utilizing a voltage controller. I bound them to the spot board this way

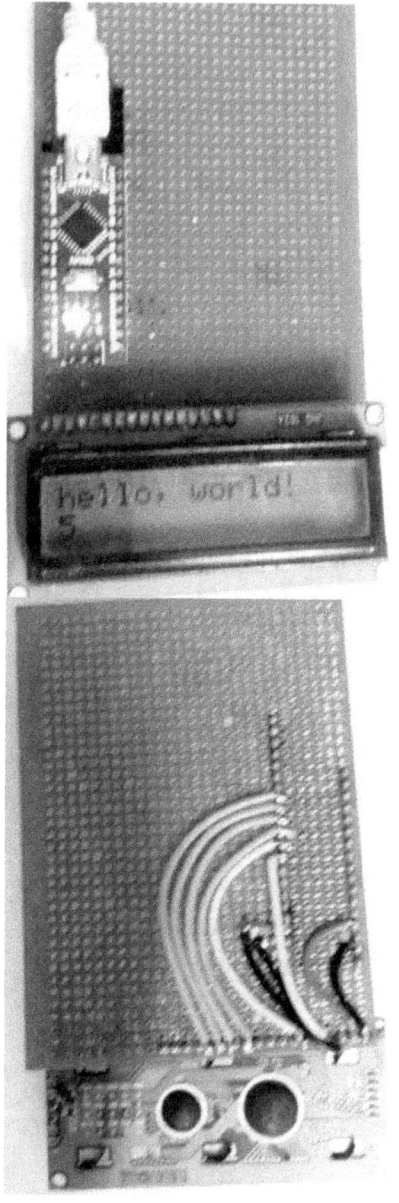

Our point is to make this RPS simple to utilize and furthermore keep the expense as low as could reasonably be expected, henceforth I have amassed it on a speck board, yet in the event that you can offered a PCB it will be incredible since we are managing high flows.

Building 0-24v 3A Variable Power Supply Circuit:

Since our Display is prepared given us a chance to begin with different circuits. From now it is fitting to continue with additional alert since we are managing AC mains and high current. Check for congruity utilizing a multimeter each time before you control you circuit.

The transformer we use is a 24V 3A transformer, this will step down our voltage (220V in India) to 24V, and we straightforwardly offer this to our extension rectifier. The scaffold rectifier should will give you (root multiple times the info voltage) 33.9V, yet don't be astonished on the off chance that you get around 27 - 30 Volts. This is a result of the Voltage drop over every diode in our extension rectifier. When we arrive at this stage we will bind it to our dab board and check our yield and utilize a terminal square with the goal that we use it as a non managed steady source whenever required.

Presently let us control the yield voltage by utilizing a high present controller like LM338K, this will be generally accessible in metal body bundle, since it needs to source high current. The schematics for variable voltage controller are demonstrated as follows.

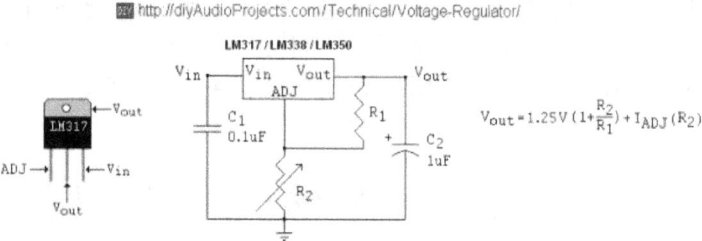

The estimation of R1 and R2 must be determined utilizing the above formulae to decide the yield voltage. You can likewise compute the resistor esteems utilizing this LM317 resistor number cruncher. For our situation we persuade R1 to be 110 ohms and R2 as 5K (POT).

When our Regulated yield is prepared we simply need to catalyst Arduino, to do this we will utilize a 7812 IC since the Arduino will just devour less current. The information Voltage of 7812 is our amended 24v DC yield from rectifier. The yield of directed 12V DC is given to the Vin stick of Arduino Nano. Try not to utilize 7805 since the most extreme info voltage of 7805 is just 24V while 7812 can withstand upto 24V. Likewise a warmth sink is required for 7812 since the differential voltage is extremely high.

The total circuit of this Variable Power Supply is demonstrated as follows,

Pursue the Schematics and weld you parts in like manner. As appeared in schematics the variable voltage of 1.5 to 24V is mapped to 0-4.5V by utilizing potential divider circuit, since our Arduino can just peruse voltages from 0-5. This variable voltage is associated with stick A0 utilizing which the yield voltage of the RPS is estimated. The last Code for the Arduino Nano is given beneath in Code Section.

When the binding work is done and the code is transferred to Arduino, our Regulated Power Supply is prepared to utilize. We can utilize any heap which works from 1.5 to 22V with a present rating of greatest 3A.

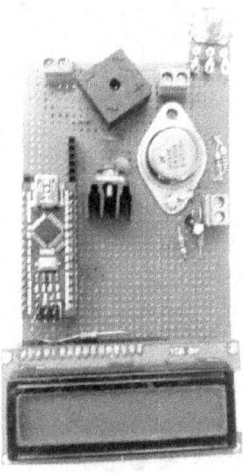

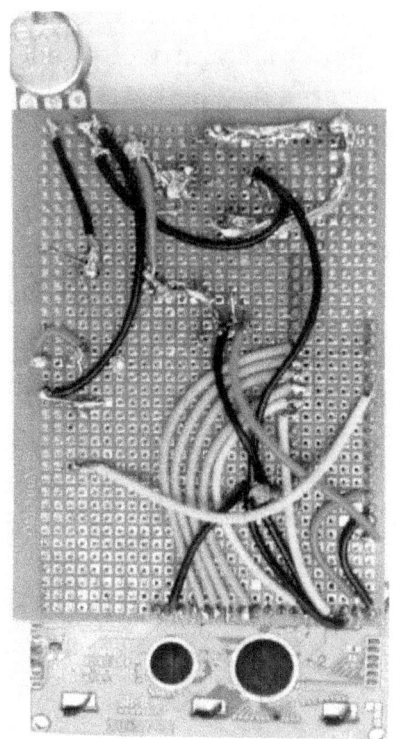

Point to be kept in mind:

1. Be cautious while binding the associations any befuddle or remissness will effectively sear your segments.

2. Conventional patches probably won't have the option to withstand 3A, this will lead in the long run liquefy your bind and cause hamper. Utilize thick copper wires or utilize more lead while interfacing the high current tracks as appeared in the image.

3. Any short out or frail patching will effectively consume your transformer windings; subsequently check for progression before fueling up the circuit. For extra wellbeing a MCB or circuit on Input side can be utilized.

4. High present voltage controllers for the most part come in metal can bundles, while utilizing them on dab board don't put segments near them as their body goes about as the yield of the redressed Voltage, further will bring about swells.

Additionally, don't bind the wire to the metal can, rather utilize a little screw as appeared in the image given underneath. Welds don't adhere to its body, and warming outcomes in harming the Regulator forever.

5. Try not to avoid any channel capacitors from the schematics, this will harm you Arduino.

6. Try not to over-burden the transformer more than 3A, stop when you hear a murmuring commotion from the transformer. It is a great idea to work between the scopes of 0 - 2.5A.

7. Confirm the yield of your 7812 preceding you interface it to your Arduino, check for overheat-

ing during first preliminary. On the off chance that warming happens it implies your Arduino is devouring progressively current, decrease the backdrop illumination of the LCD to comprehend this.

Upgrade:

The Regulated Power Supply (RPS) that is posted above have hardly any issue with the precision because of the commotion present in the yield signal. This sort of commotion is regular in situations where an ADC is utilized, a straightforward answer for it is to utilize a low pass channel like RC channel. Since our circuited Dot board has both AC and DC in its trails, the commotion will be high than that of different circuits. Henceforth an estimation of R= 5.2K and C= 100uf is used to sift through the commotion in our sign.

Likewise a present sensor ACS712 is added to our circuit to quantify the yield current of the RPS. The underneath dissident tells the best way to interface the Sensor to the to Arduino Board.

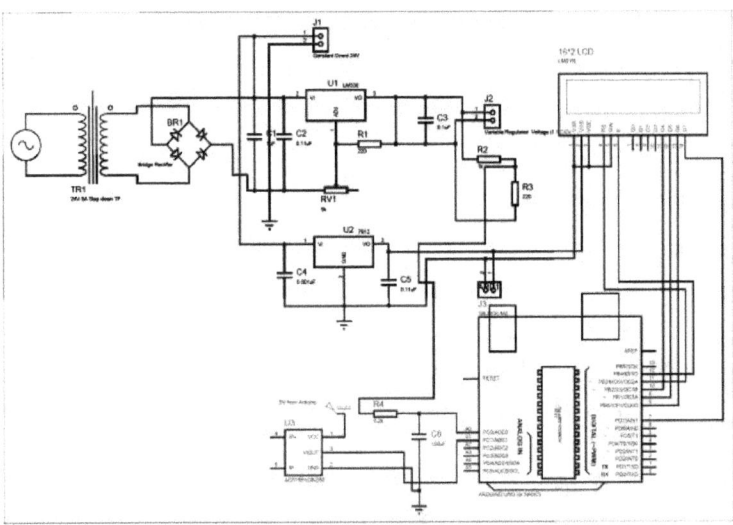

Code

```
#include <LiquidCrystal.h>
// initialize the library with the numbers of the interface pins
LiquidCrystal lcd(7, 8, 9, 10, 11, 12);
void setup()
{
 Serial.begin(9600);
 // set up the LCD's number of columns and rows:
 lcd.begin(16, 2);
 // Print a message to the LCD.
 lcd.setCursor(0, 0);
 lcd.print("RPS");
 lcd.setCursor(0, 1);
 lcd.print("-Hello World");
 delay(2000);
```

```
lcd.clear();
lcd.setCursor(0, 0);
lcd.print("Voltage = ");
}
int voltage;
void loop()
{
 int A1 = analogRead(A0);
 voltage = map(A1,0,1024,0,22);
 Serial.println(voltage);
 lcd.setCursor(10,0);
 lcd.print(voltage);
 delay(1000);
}
```

◆ ◆ ◆

3. ENTRYWAY ALARM UTILIZING ARDUINO AND ULTRASONIC SENSOR

Security has consistently been a significant worry for us all and there are numerous Hi tech and IoT based security and reconnaissance framework are accessible in the market. Gatecrasher or Burglar Alarm is one of the work of art and well known undertaking among the Electronics understudies and specialists. We have likewise fabricated numerous Burglar Alarms dependent on different advances:

- Laser Security Alarm Circuit

- IR Based Security Alarm

- Robber Alarm utilizing PIR

- GSM Based Security System

Today we are including one greater Security Alarm in our rundown which depends on Ultrasonic Sensor. This Arduino Controlled Door alert can be introduced close to the entryway to identify the nearness of anyone at the entryway. At whatever point someone comes in the scope of Ultrasonic sensor, ringer starts blaring. You can alter the sensor recognition range as indicated by your entryway. This framework can likewise fill the need of Motion Detector.

Required Components:

- Breadboard
- Buzzer
- Ultrasonic Sensor
- Jumper Wires
- Arduino Mega (any model)
- USB link for Arduino or 12v, 1A connector.

Ultrasonic Sensor Module:

Ultrasonic sensor HC-SR04 is utilized here to identify the existences of any individual at the entryway. The sensor module comprises of ultrasonic transmitter, collector and the control circuit. Ultrasonic Sensor comprises of two round eyes out of which one is utilized to transmit the ultrasonic wave and the other to get it.

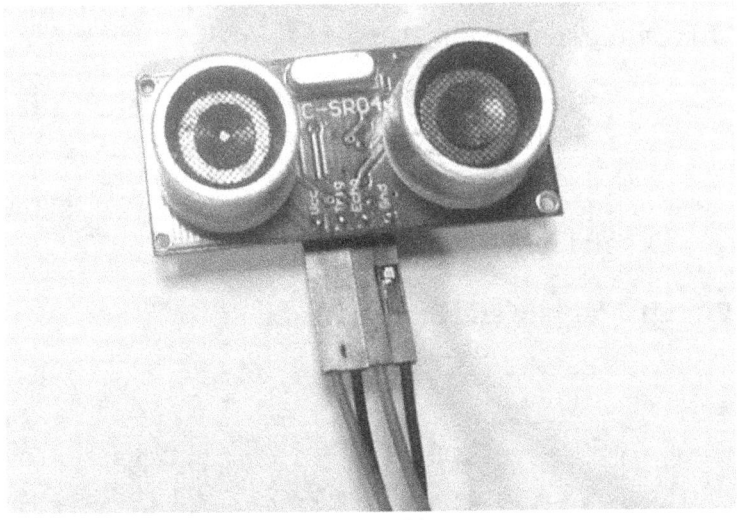

We can compute the separation of the article dependent on the time taken by ultrasonic wave to return back to the sensor. Since the time and speed of sound is realized we can ascertain the separation by the accompanying formulae.

- Separation = (Time x Speed of Sound)/2

The worth is isolated by two since the wave goes ahead and in reverse covering a similar separation. In any case, in this undertaking we have utilized NewPing.h library, and this library deals with this estimation and we simply need to utilize some catchphrases, clarification is given in programing area underneath.

Check the beneath venture to quantify the separation of any item and to appropriately comprehend the Ultrasonic sensor working:

- Arduino Based Distance Measurement utilizing Ultrasonic Sensor

- Separation Measurement utilizing HC-SR04 and AVR Microcontroller

Circuit Diagram and Explanation:

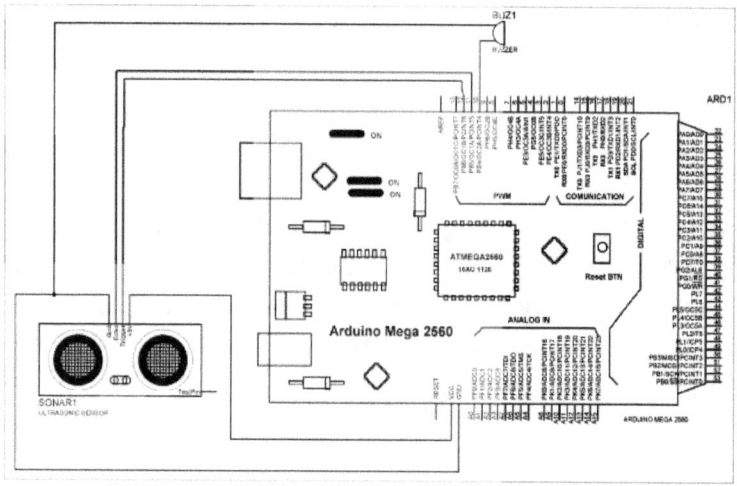

Circuit associations for this Ultrasonic Alarm are straightforward. Trigger stick of ultrasonic sensor is associated with stick no. 12 of Arduino and Echo stick of sensor is associated with stick no 11 of Arduino. Vcc of sensor is associated with 5V stick of Arduino and GND of sensor is associated with GND of Arduino. One stick of ringer is associated with GND of

Arduino and the other stick is associated with eighth stick of Arduino.

Working Explanation:

Working this Arduino Door Alarm is exceptionally simple. At whatever point anybody comes in the way/scope of Ultrasonic Sensor, microcontroller recognizes the separation of item from the sensor and if the article is in the characterized range, it sends the High sign to the ringer and bell starts signaling.

You can test the circuit by placing anything before sensor inside the range. This capacity of Ultrasonic Sensor can likewise be utilized to manufacture Obstacle Avoiding Robot.

Programming Explanation:

In this undertaking we have utilized NewPing.h Library for Ultrasonic sensor, created by Tim Eckel. In spite of the fact that we can utilize Ultrasonic sensor without this library as we did in our past venture, however this Library gives numerous great highlights to Ultrasonic sensor and it turns out to be anything but difficult to Code for ultrasonic sensor utilizing this library. We can utilize ultrasonic sensor's capacities effectively utilizing this library without composing an excessive number of lines of code; it resembles different libraries which are utilized to deal with the multifaceted nature at lower level.

```
#include <NewPing.h>
```

You can check every one of the highlights, uses and test codes of this Library by following this connection. Likewise check the authority Arduino page of this Library.

Most recent arrival of Library can be downloaded from the above given connection. Further, we have downloaded the Library from beneath connect, which is adjusted for Teensy:

https://github.com/PaulStoffregen/NewPing

You should initially test the sensor by consuming Example Codes given at its page. We have likewise utilized the Example Codes for our task and changed them as indicated by our Door Alarm Project.

Trigger stick is associated with the Pin 12 of Arduino and Echo stick is associated with stick 11 of Arduino. MAX_DISTANCE implies that the separation up to which the sensor can recognize the deterrent is 500 cm or 5m.

```
#define TRIGGER_PIN 12

#define ECHO_PIN   11

#define MAX_DISTANCE 500
```

Beneath line expresses the Baud Rate at which the information is sent to the Arduino sequential port from ultrasonic sensor.

```
Serial.begin(115200);
```

Stick no 10 is designed as yield stick and is associated with ringer. Other stick of signal is associated with GND of Arduino.

```
pinMode(10, OUTPUT);
```

In void echoCheck() work, sonar.ping_result/ US_ROUNDTRIP_CM is utilized to ascertain the separation of impediment from the sensor. banner is utilized to execute the signal when hindrance is in 50 cm go from the ultrasonic sensor. You can change this 'separation' as indicated by your prerequisite or your entryway size.

```
if((sonar.ping_result / US_ROUNDTRIP_CM) < 50)

    flag = 1;

  else if((sonar.ping_result / US_ROUNDTRIP_CM)
> 50)

    flag = 0;
```

The Code is very much remarked by the Author of NewPing.h library and can be effectively comprehended. Further you can check the library page itself to show signs of improvement comprehension of it and can utilize this library to make complex task utilizing ultrasonic sensor. Full code for this Door Alarm task is given underneath.

Basically Ultrasonic sensor is utilized to gauge good ways from any item, however here we can see that it very well may be utilized as Security caution or Door alert with Arduino. In like manner we can make numerous valuable tasks utilizing this like: Automatic Water Level Indicator and Controller utilizing Arduino

Code

```
#include <NewPing.h>
#define TRIGGER_PIN 12 // Arduino pin tied to trigger pin on ping sensor.
```

```
#define ECHO_PIN    11 // Arduino pin tied to echo
pin on ping sensor.
#define MAX_DISTANCE 500 // Maximum distance
we want to ping for (in centimeters). Maximum sen-
sor distance is rated at 400-500cm.

NewPing sonar(TRIGGER_PIN, ECHO_PIN, MAX_DIS-
TANCE); // NewPing setup of pins and maximum dis-
tance.

unsigned int pingSpeed = 50; // How frequently are
we going to send out a ping (in milliseconds). 50ms
would be 20 times a second.
unsigned long pingTimer;
int flag = 0; // Holds the next ping time.

void setup() {
   Serial.begin(115200); // Open serial monitor at
115200 baud to see ping results.
 pingTimer = millis();
 pinMode(10, OUTPUT); // Start now.
 // Start now.
}
void loop() {
 // Notice how there's no delays in this sketch to
allow you to do other processing in-line while doing
distance pings.
 if (millis() >= pingTimer) {  // pingSpeed millisec-
onds since last ping, do another ping.
 pingTimer += pingSpeed;   // Set the next ping time.
 sonar.ping_timer(echoCheck); // Send out the ping,
calls "echoCheck" function every 24uS where you can
check the ping status.
```

```
}
if(flag == 1)
{
 digitalWrite(10, HIGH);
 delay(500);
 digitalWrite(10, LOW);
 delay(500);
 digitalWrite(10, HIGH);
 delay(500);
 digitalWrite(10, LOW);
 delay(500);
}
else
{
 digitalWrite(10, LOW);
}
}
void echoCheck() { // Timer2 interrupt calls this
function every 24uS where you can check the ping
status.
 if(sonar.check_timer()) { // This is how you check to
see if the ping was received.
  // Here's where you can add code.
  Serial.print("Ping: ");
                 Serial.print(sonar.ping_result   /
US_ROUNDTRIP_CM); // Ping returned, uS result in
ping_result,      convert      to      cm      with
US_ROUNDTRIP_CM.
  Serial.println("cm");
  if((sonar.ping_result / US_ROUNDTRIP_CM) < 50)
```

```
  flag = 1;
  else if ((sonar.ping_result / US_ROUNDTRIP_CM) >
50)
  flag = 0;
 }
}
```

◆ ◆ ◆

4. DIY SMART VACUUM CLEANING ROBOT UTILIZING ARDUINO

Greetings folks, would you say you are a beginner to the universe of mechanical autonomy or electronic? Or on the other hand Are you searching for a straight-forward yet amazing venture to make your companions and educators intrigued? At that point this is the spot.

In this venture we will utilize the intensity of Embedded Systems and Electronics to make our own robot which could help us in keeping our home or work place slick and clean. This robot is basic four wheeled Vacuum Cleaner which could insightfully maintain a strategic distance from deterrents and vacuum the floor simultaneously. The thought is motivated by the acclaimed vacuum cleaner Robot Roomba which is appeared in the picture beneath.

Our Idea is to make a straightforward robot directly from the scratch which can naturally maintain a strategic distance from the obstructions while cleaning the floor. Trust me individuals it's entertaining!!

Required Material and Components:

Alright so now we have the Idea of our Automatic Vacuum Cleaner Robot as a primary concern and we comprehend what we are doing. So how about we look where we should begin our execution. So as to fabricate a robot of our thought we would initially need to choose the accompanying:

- Microcontroller type
- Motors required
- Sensors required
- Battery capacity
- Robot chassis material

Presently, gives us a chance to choose every one of the previously mentioned focuses. Along these lines it will be useful for you to fabricate this home cleaning robot as well as whatever other robots which strikes your creative mind.

Microcontroller Type:

Choosing the Microcontroller is a significant assignment, as this controller will go about as the cerebrum of your robot. The majority of the DIY undertakings are made around Arduino and Raspberry Pi, however doesn't need to be the equivalent. There is no particular Microcontroller that you can wear down. Everything relies on the necessity and cost.

Like a Tablet can't be planned on 8 piece Microcontroller and there is no value of utilizing ARM cortex m4 to structure an electronic number cruncher.

Microcontroller choice absolutely relies on the prerequisites of the item:

1. From the start specialized prerequisites are dis-

tinguished like number of I/O pins required, streak size, number/sort of correspondence conventions, any extraordinary highlights and so forth.

2. At that point rundown of controllers are chosen according to the specialized necessities. This rundown contains controllers from various makers. Numerous application explicit controllers are accessible.

3. At that point a controller is concluded dependent on cost, accessibility and backing from producer.

In case you would favor not to do part of truly difficult work and simply need to get acquainted with the nuts and bolts of microcontrollers and afterward get profound into it, at that point you can pick Arduino. In this venture we will utilize an Arduino. We have recently made numerous kinds of Robots utilizing Arduino:

- DTMF Controlled Robot utilizing Arduino

- Line Follower Robot utilizing Arduino

- PC Controlled Robot utilizing Arduino

- WiFi Controlled Robot utilizing Arduino

- Accelerometer Based Hand Motion Controlled Robot utilizing Arduino

- Bluetooth Controlled Toy Car utilizing Arduino

Sensors Required:

There are a ton of sensors accessible in the market each having its very own use. Each robot gets input by means of a sensor, they go about as the tactile organs for the Robot. For our situation our robot ought to have the option to recognize hindrances and maintain a strategic distance from them.

There a ton of other cool sensor which we will use in our future undertakings, however now let us remain concentrated on IR sensor and US (Ultrasonic sensor) as these two folks will give the vision to our robo-vehicle. Look at the working of IR sensor here. Beneath demonstrating pictures of IR sensor Module and Ultrasonic Sensor:

Ultrasonic Sensor comprises of two round eyes out of which one is utilized to transmit the US signal and the other to get the US beams. The time taken by the beams to get transmitted and got back is determined by the microcontroller. Presently, since the time and speed of sound is realized we can figure the separation by the accompanying formulae.

- Separation = Time x Speed of Sound partitioned by 2

The worth is separated by two since the beam goes ahead and in reverse covering a similar separation. Itemized clarification of utilizing Ultrasonic sensor is given here.

Engines required:

There are a considerable amount an engines utilized

in the field of apply autonomy the most utilized ones are the Stepper and Servo engine. Since this undertaking doesn't have any convoluted actuators or rotating encoder we will utilize a typical PMDC Motor. Be that as it may, our battery is somewhat massive and overwhelming consequently we utilize four engines to drive our robot each of the four being the equivalent PMDC engines. Yet, it is prudent to set into stepper and servo engines once you get settled with PMDC engines.

Robot body material:

As an understudy or specialist the most troublesome part while making a robot is to set up the suspension of our robot. The issue is with the accessibility of instruments and material. The best material for this venture will be Acrylic, yet it requires drillers and different instruments to work with it. Subsequently wood is picked that everybody can take a shot at it easily.

This issue has completely disappeared from the field after the presentation of the 3D printers. I am wanting to 3D print parts some time or another and update you individuals with the equivalent. So until further notice how about we utilize wooden sheets to fabricate our robot.

Battery limit:

Choosing the battery limit ought to be our last piece of work since it simply relies upon your frame and engines. Here our battery should drive a vacuum cleaner which draws around 3-5A and four PMDC engines. Henceforth we will require an overwhelming battery. I have picked 12V 20Ah SLAB (Sealed lead corrosive battery) and its quite cumbersome causing our robot to get four PMDC engines to pull this massive person.

Since we have chosen all our necessary segments lets show them down

- Wooden sheets for suspension

- IR as well as US sensors

- Vacuum cleaner which runs on DC current

- Arduino Uno

- 12V 20Ah battery

- Engine driver IC (L293D)

- Working apparatuses

- Interfacing wires

- Energetic vitality to learn and work.

The majority of our segments are canvassed in the portrayal above, I will clarify the left outs beneath.

DC vacuum more clean:

Since our robot runs on a 12V 20Ah DC framework. Our vacuum ought to likewise be a 12V DC vacuum more clean. In the event that you are confounded on where to get one, at that point you can visit eBay or Amazon for vehicle cleaning vacuum cleaners.

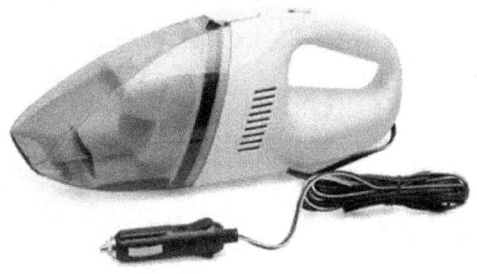

We will utilize equivalent to appeared in above picture.

Engine driver (L293D):

An engine driver is a halfway module among Arduino and the Motor. This is on the grounds that Arduino microcontroller won't have the option to supply the current required for the engine to work it and can simply supply 40mA, consequently drawing increasingly current will harm the controller for all time. So we trigger the engine driver which thus controls the engine.

We will utilize L293D Motor Driver IC which will have the option to supply up to 1A, thus this driver will get the data from Arduino as well as make the engine function as wanted.

That is it!! I have given the greater part of the essential data however before we start fabricating the robot it is prescribed to experience the datasheet of L293D and Arduino.

Building and Testing the Robot:

The Vacuum Cleaner is the most essential part in arrangement of Robot. It must be set at tilted edge as appeared in the image, with the goal that it can give legitimate vacuum activity. The vacuum cleaner isn't constrained by the Arduino. When you control on the robot the vacuum is likewise turned on.

One tiring procedure of building our Robot is the wooden works. We need to cut our wood and drill a few openings to put the sensors and vacuum more clean.

It is prescribed to Test Ride your Robot with the accompanying code once you mastermind the Motor and Motor driver, before interfacing the Sensors.

```
void setup()

{

Serial.begin(9600);

pinMode(9,OUTPUT);

pinMode(10,OUTPUT);

pinMode(11,OUTPUT);

pinMode(12,OUTPUT);

}

void loop()

{

delay(1000);

Serial.print("forward");
```

```
digitalWrite(9,HIGH);

digitalWrite(10,LOW);

digitalWrite(11,HIGH);

digitalWrite(12,LOW);

delay(500);

Serial.print("backward");

digitalWrite(9,LOW);

digitalWrite(10,HIGH);

digitalWrite(11,LOW);

digitalWrite(12,HIGH);

}
```

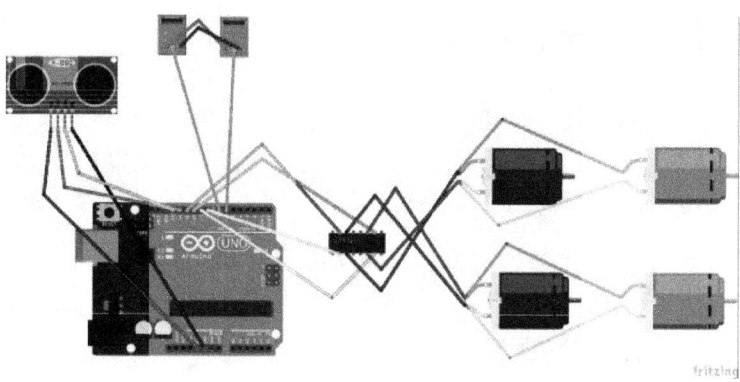

In case everything works fine, at that point you can interface the sensors with Arduino as appeared in Circuit Diagram and utilize the Full Code given toward the end. As should be obvious I have mounted a Ultrasonic sensor to the front as well as two IR sensors on both the side of the robot. The warmth sink is fitted on to the L293D just in the event that the IC warms up quick.

You can likewise include barely any additional parts like this one

This is a Sweeping Arrangement can be put on the two parts of the bargains part that will push the residue at the edges into the suction region.

Further, you likewise have an alternative of making a Smaller Version of this Vacuum Cleaning Robot like this

This littler Robot is made on cardboard and runs on ATMega16 improvement board. The vacuum cleaner part was finished by utilizing a BLDC fan and encased in a case. You can embrace this in the event that you have to keep your spending low. This thought additionally works yet it's not productive.

Circuit Diagram:

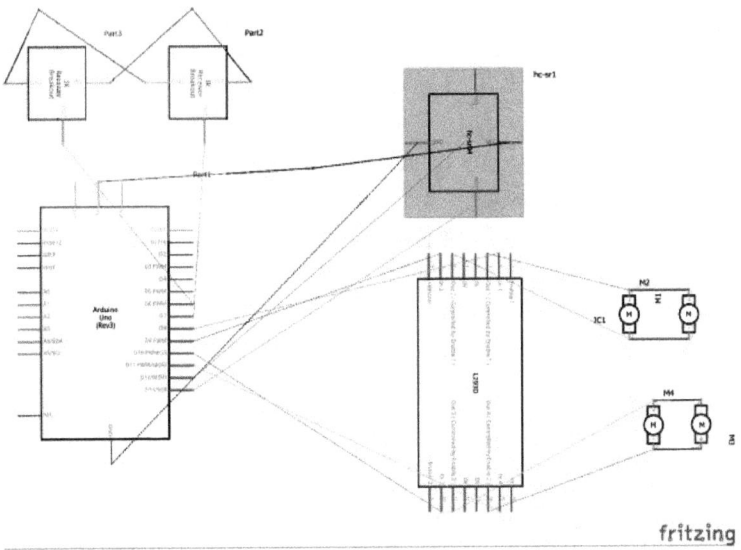

fritzing

This Vacuum Cleaner Robot code can be found in the Code Segment underneath. When the association is done and program is dumped into Arduino, your robot is prepared to get without hesitation. In the event that you have to see this robot in real life.

Further, I am additionally intending to totally 3D Printed the parts in its next adaptation. I am additionally going to include hardly any cool highlights and complex calculations with the goal that it covers the entire rug region and simple to deal with and minimized in size.

Code

#define trigPin 12
#define echoPin 13

```
#define ir1 7
#define ir2 6
void setup()
{
 Serial.begin(9600);
 pinMode(8,OUTPUT);
 pinMode(9,OUTPUT);
 pinMode(10,OUTPUT);
 pinMode(11,OUTPUT);
 pinMode(trigPin, OUTPUT);
 pinMode(echoPin, INPUT);
 pinMode(ir1, INPUT);
 pinMode(ir2,INPUT);
}
void loop()
{
 int duration, distance;
 int flag,val1,val2;
 val1=digitalRead(ir1);
 val2=digitalRead(ir2);
 Serial.println(val1);
 Serial.println(val2);
 digitalWrite(trigPin, HIGH);
 delayMicroseconds(1000);
 digitalWrite(trigPin, LOW);
 duration = pulseIn(echoPin, HIGH);
 distance = (duration/2) / 29.1;
 if(distance >= 200 || distance <= 0){
  Serial.println("Out of range");
 }
```

```
 else {
  Serial.print(distance);
  Serial.println(" cm");
 }
 delay(500);
if(distance >=20)
{
 delay(100);
 Serial.println("forward");
 digitalWrite(8,HIGH);
 digitalWrite(9,LOW);
 digitalWrite(10,HIGH);
 digitalWrite(11,LOW);
 delay(150);
 Serial.println("STOP");
 digitalWrite(8,LOW);
 digitalWrite(9,LOW);
 digitalWrite(10,LOW);
 digitalWrite(11,LOW);
}
if(distance<=20)
{
 if(val1==1 )
 {
  Serial.print("turn right");
  digitalWrite(8,LOW);
  digitalWrite(9,LOW);
  digitalWrite(10,HIGH);
  digitalWrite(11,LOW);
  delay(2000);
```

```
}
if(val2==1)
{
 Serial.print("turn left");
 digitalWrite(8,HIGH);
 digitalWrite(9,LOW);
 digitalWrite(10,LOW);
 digitalWrite(11,LOW);
 delay(500);
}
}
if(distance<=10)
{
 Serial.print("rearrange back");
 digitalWrite(8,LOW);
 digitalWrite(9,HIGH);
 digitalWrite(10,LOW);
 digitalWrite(11,HIGH);
 delay(1000);
 Serial.print("rearranged left");
 digitalWrite(8,LOW);
 digitalWrite(9,LOW);
 digitalWrite(10,HIGH);
 digitalWrite(11,LOW);
 delay(100);
}
if(distance<=20)
{
 Serial.print("Algorithum");
 Serial.print("free right");
```

```
digitalWrite(8,HIGH);
digitalWrite(9,LOW);
digitalWrite(10,LOW);
digitalWrite(11,LOW);
if(val2==0)
{
Serial.print("smart adjust");
digitalWrite(8,LOW);
digitalWrite(9,HIGH);
digitalWrite(10,LOW);
digitalWrite(11,LOW);
delay(500);
}
}
}
```

◆ ◆ ◆

5. ADVANCED MOBILE PHONE CONTROLLED FM RADIO UTILIZING ARDUINO AND PROCESSING

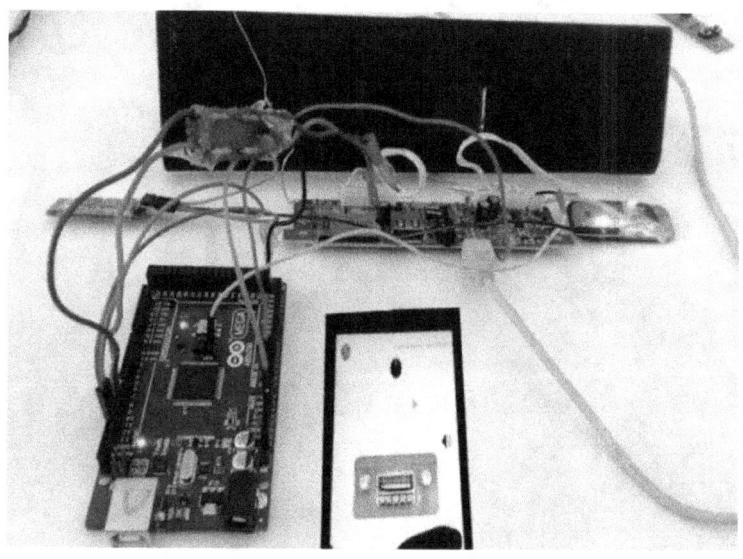

In this task we will utilize a current FM radio which went fix quite a while prior, to change over it into a Smart Wireless FM Radio controlled utilizing Phone, with the assistance of Arduino and Processing.

We can change over any physically worked electronic gadget into a Smart Device utilizing a similar method. Each electronic gadget works with the assistance of sign. These sign may be as far as voltages or flows. The sign can either be activated physically with the assistance of client cooperation straightforwardly or with the assistance of a remote gadget.

Before the finish of this task we will have the option to change over the large majority of our normal electronic gadgets, similar to a Radio which chips away at catches, into a Smart Wireless Gadget which can be constrained by Smart telephone over Bluetooth. To accomplish this we should accomplish two principle things.

1. Foresee how the sign are produced in the current mechanical catch framework.

2. Discover an approach to trigger a similar sign with assistance of a little extra circuit.

Along these lines, Let's begin...

Components Required:

For this task an old or unused electronic gadget like a radio, TV, CD player, or Home auditorium can be chosen. The genuine parts may change dependent on the gadget you select. Yet, to make it remote we would require a microcontroller which is an Arduino here as well as a remote medium which is a HC-05 Bluetooth module.

Reverse Engineering:

Alright, so now I have chosen an old FM radio player which quit working quite a while back. What's more, when I opened it I found that the catches on it have quit working. This will be an ideal gadget for us to work since we won't require the catches any longer as we are going to make it remote totally. The underneath picture shows the Radio which I opened.

This was the catch arrangement of my radio (above picture). As should be obvious there are eight catches from which the radio takes input. You can likewise see that there are eight resistors on the board. What would you be able to finish up from this... ? Indeed every resister is associated with a switch. Presently how about we investigate the posterior of the board:

You can follow out the association with the assistance of the PCB tracks, however in case you are as yet confounded you can utilize your millimeter in network more and make sense of the circuit. This board has three terminals (hovered in red) which offers sign to the primary FM radio board. These pins were set apart as S1, S2, and 1.7V. This implies consistent voltage of 1.7 Volts is sent structure the primary board to this board and as the client presses any catch, there will be a voltage drop over the relating resistor and through the pins S1 and S2 a variable voltage will be sent back. This is the manner by which a large portion of the catches in our electronic gadgets work. Presently since we have made sense of how it functioned, we should make it remote.

Working Explanation:

So now to make it remote we simply need to give a voltage between 0 - 1.7V over the S1 and ground out the principle board. There are hardly any ways, utilizing which you can emulate these catch arrangement utilizing a microcontroller.

We can utilize a Digital potentiometer and cause it to give the obstruction on the board as customized and when required. Be that as it may, this will make things confounded and expensive as working with Digipot requires SPI and Digipots are exorbitant.

We can likewise utilize a transistor resistor arrange in which every resistor of various qualities are enacted by a transistor which thusly is constrained by the microcontroller itself. In any case, again to do this for eight fastens the circuit will get confounded.

The straightforward method for doing this is to legitimately produce the necessary variable voltage from the microcontroller and feed it to the sign pins. Tragically, Arduino just has ADC and doesn't have a DAC. Be that as it may, fortunately we have PWM in Arduino. This PWM can be made to go about as a variable voltage with the assistance of a basic RC Low Pass Filter.

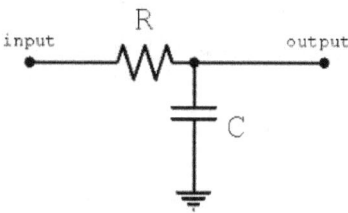

A low pass channel is appeared over, the key part here is the capacitor which will ground the whole throbbing sign and an unadulterated DC is sent as yield. So the PWM signals from the Arduino must be sent through a low pass channel and afterward given to the sign leading body of the FM radio.

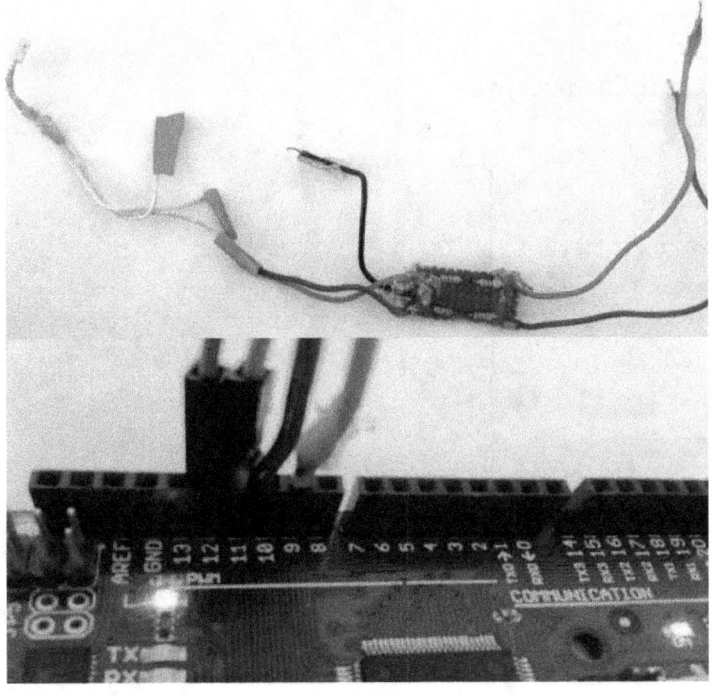

The circuit is anything but difficult to expand on a dab board as appeared previously. Here the dark wire is for ground as well as the blue as well as green wires on the left will be sent to our fm sheets S1(Green) as well as S2(blue), as well as the wires to the correct will get PWM signals from Arduino's Pin 9 and 10 (see picture above) as well as go to the FM board through a Low pass channel. The Bluetooth module uses pins 11 and 12 as Rx and TX.

Presently we can produce PWM signals from 0 volt to 1.7 volt and discover how our Radio acts for various voltage levels. The subsequent stage is to make this thing remote.

Circuit Connections:

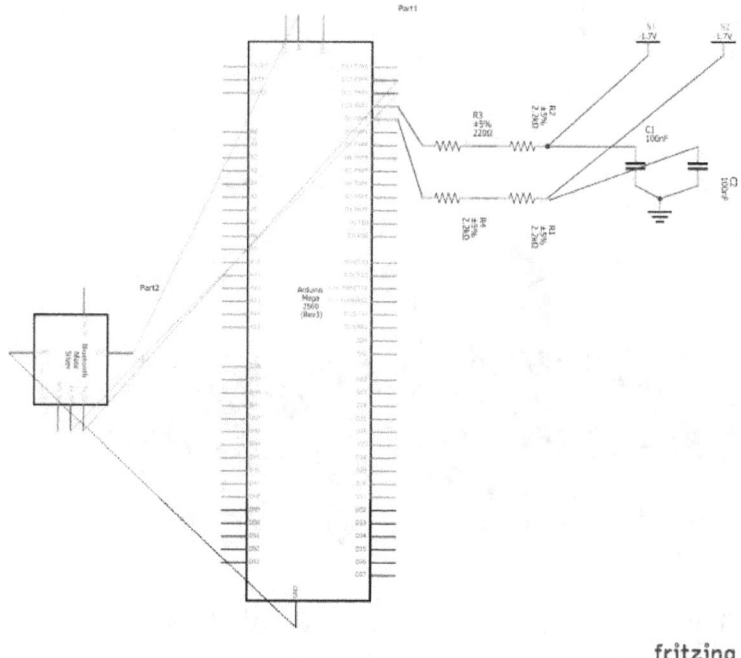

This schematic shows the whole arrangement of Low Pass Filter and HC-05 Bluetooth Module associated with Arduino Mega for Bluetooth Controlled FM Radio.

Arduino Program:

Program for the Arduino is given in the Code segment beneath. You can likewise test the Variable Voltage go for your electronic gadget by utilizing this program here.

Check this Article to arranging Bluetooth Terminal App on Arduino. When we are certain with its working, we can hop into making our own Android application.

Using Processing to Make Android App:

It is cool to make our gadget remote, yet we can likewise add some close to home touch to our gadget by making our very own Android application. We can control the gadget on programmed booked occasions or control it dependent on your wake up alerts. Even you can make your radio play your preferred channel when you return home. Creative mind is your point of confinement here. Be that as it may, for the present we will make a straightforward UI utilizing handling, this application will just have barely any catches utilizing which you can control your FM radio.

Preparing is open source programming which is utilized by craftsmen for Graphics structuring. This product is utilized to create programming and Android applications.

The Processing Code for the Android App to control this Wireless FM Radio is given here:

- Android App Processing Code to control the FM Radio

First we assembled this application on PC in JAVA mode, to test it appropriately, here is the Processing Code for the equivalent. Right click on it and snap on 'Spare interface as..' to download the code record. At that point open the document in 'Handling' programming and snap on 'Run' catch to check how it will look in the Phone. You have to introduce 'Handling' programming to open *.pde documents.

When we have tried out App in JAVA mode we can undoubtedly change over it into Android Mode by changing to Android tab on the upper right corner of the Processing window. So as to make our Android Phone turn on its Bluetooth and associate with our HC-05 module naturally, we have to add the accompanying codes to our current Java program to make it an Android App. We have just given the full Android Code in above connection, so you can straightforwardly utilize it.

The following are some Header records to empower Bluetooth capacities:

```
import android.content.Intent;

import android.os.Bundle;

import ketai.net.bluetooth.*;
```

```
import ketai.ui.*;
```

```
import ketai.net.*;
```

```
import android.bluetooth.BluetoothAdapter;
```

```
import android.view.KeyEvent;
```

Beneath lines speaks with our telephones Bluetooth connector utilizing Ketai library and we name our connector as bt.

```
BluetoothAdapter bluetooth = BluetoothAdapter.getDefaultAdapter();

KetaiBluetooth bt;
```

Beneath some portion of the code will trigger a solicitation to the client requesting that they Turn on the Bluetooth on application start up.

```
//To start BT on start*********

void onCreate(Bundle savedInstanceState) {

super.onCreate(savedInstanceState);

bt = new KetaiBluetooth(this);
```

```
}

void onActivityResult(int requestCode, int result-
Code, Intent data) {

  bt.onActivityResult(requestCode,      resultCode,
data);

}

//*********
```

Here we train our Android App to which Blue-tooth gadget we need to get associated with. The line bt.connectToDeviceByName(selection); expect a gadget name from our arrangement work. Since our Bluetooth gadget is named as 'HC-05', beneath line is included the arrangement. This name will vary dependent on your Bluetooth modules name.

```
//To select bluetooth device*********

void onKetaiListSelection(KetaiList klist)

{

String selection = klist.getSelection();

bt.connectToDeviceByName(selection);
```

```
//dispose of list for now

klist = null;

}

//**********
```

```
bt.connectToDeviceByName("HC-05");
```

Possibly you can do these adjustments in Processing Code for PC (Java mode) or can legitimately utilize our Android Processing code given in above connection. At that point straightforwardly interface your telephone to your PC utilizing the information link and empower USB investigating on your telephone. Presently click on the Play button on the handling window in PC, the application will be legitimately introduced on your Android Phone and will be propelled consequently. It's that simple, so feel free to give it a shot.

The beneath picture speaks to our Android Application UI alongside its coding window. Run the Code in Android Phone just as in PC.

That is it we have transformed our old FM radio into a remote present day device that can be constrained by our Android Application.

Code

```
int outvalue =0;;
const int GPWM = 9;
const int BPWM = 10;
int invalue;
```

```
#include <SoftwareSerial.h>// import the serial library
SoftwareSerial Genotronex(11, 12); // TX, RX
int BluetoothData; // the data given from Computer
void setup()
{
Serial.begin(57600);
Genotronex.begin(9600);
Serial.println("Enter Value to write (60-195)");
TCCR2B = (TCCR2B & 0xF8) | 0x01;// timer frequency
is 4khz
}
void loop()
{
if (Genotronex.available())
{
BluetoothData=Genotronex.read();
 invalue= BluetoothData;
if (invalue == 'u')
{
 Serial.println("Volume up");
 outvalue= 45;
 GreenDigibutton();
}
if (invalue == 'd')
{
 Serial.println("Volume down");
 outvalue= 35;
 GreenDigibutton();
```

```
}
if(invalue == 'm')
{
 Serial.println("Mode change");
 outvalue= 65;
 GreenDigibutton();
}
if(invalue == 's')
{
 Serial.println("Stop");
 outvalue= 75;
 GreenDigibutton();
}
if(invalue == 'p')
{
 Serial.println("Prev. Channel");
 outvalue= 35;
 BlueDigibutton();
}
if(invalue == 'n')
{
 Serial.println("Next Channel");
 outvalue= 45;
 BlueDigibutton();
}


   }


   }
```

```
void GreenDigibutton()
{
 analogWrite(GPWM, outvalue);
 delay(200);
 analogWrite(GPWM, 0);
  delay(200);
   Serial.println("DONE");
}
void BlueDigibutton()
{
 analogWrite(BPWM, outvalue);
 delay(200);
 analogWrite(BPWM, 0);
  delay(200);
   Serial.println("DONE");
}
```

❖ ❖ ❖

6. THE MOST EFFECTIVE METHOD TO USE NEOPIXEL LED STRIP WITH ARDUINO AND TFT LCD

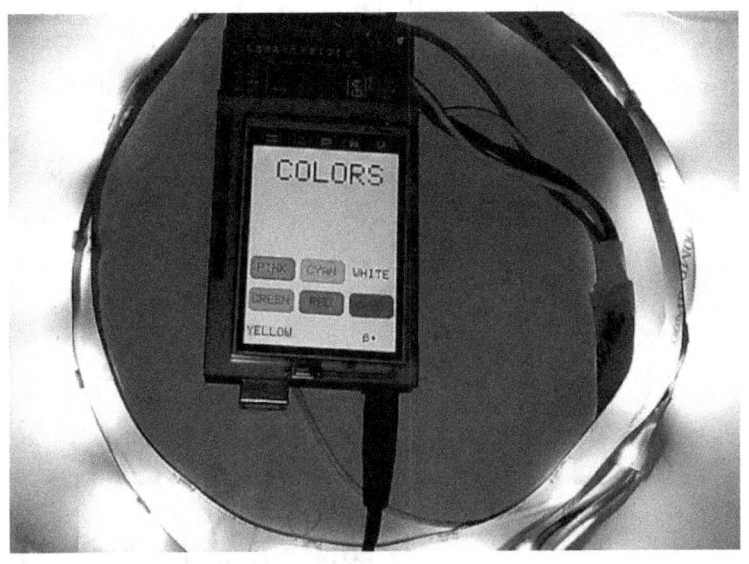

Any shading is comprised of three hues: Red, Green as well as Blue, utilizing a basic RGB LED you can create any shading. Be that as it may, the confinement of RGB LED is that it has 3 separate LEDs inside it and requires three Pins of any microcontroller to work one

RGB LED. So it is absurd to expect to interface many LEDs with one microcontroller.

To defeat this issue Adafruit has made NeoPixel LED Strip. It requires just three pins to drive a few RGB NeoPixel LEDs. Two pins are for power and ground and one Pin is for Data In (DI). Information IN stick is utilized to address and control the various LEDs in the strip with their shading, brilliance and so on. Be that as it may, it requires a Microcontroller to run NeoPixels. Arduino is regularly utilized with Neo-Pixel, so today we will figure out how to Interface NeoPixel LEDs with Arduino. You can become familiar with NeoPixels at AdaFruit.

Here in this task we are Controlling NeoPixel LED utilizing Arduino and TFT LCD contact Screen. We have made 7 touch catches of various hues on 2.4 inch TFT LCD and when we tap the catch of certain shading on the LCD, the NeoPixel LED strip lights up with a similar shade of that catch. Here we have utilized NeoPixel Digital RGB LED segment of 30 LEDs.

NeoPixel RGB LED can be lit up in any shading thus we can include more fastens the LCD to gleam the LED in more hues on tapping on those catches. Other lovely impacts and examples can likewise be included utilizing Coding. You can fabricate a full Arduino controlled Decoration System utilizing NEO Pixel LEDs and can control this framework by LCD lying close to you.

Required components:

- Arduino Mega or some other Arduino model

- NeoPixel RGB LED Strip

- 2.4 inch TFT LCD Shield with SPFD5408 controller

- USB Cable or 12 V 1A connector

- Associating Wires

Circuit Connections:

To associate NeoPixels Strip to Arduino Mega basically interface Arduino 5V stick to NeoPixel's 5V stick and Mega's GND to NeoPixel's GND and afterward interface NeoPixel DI stick (information in) to Digital Pin no 36 of Arduino Mega. Cautiously mount the TFT LCD Touch Shield over Arduino to such an extent that GND of MEGA lies underneath GND of LCD and 5V stick of Arduino interfaces with 5V stick of LCD.

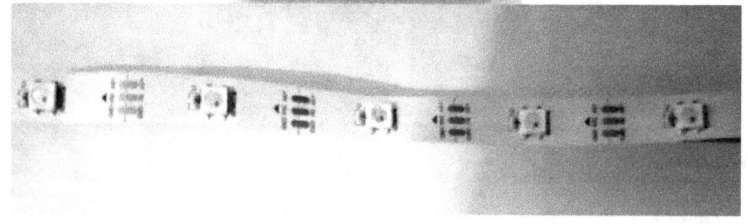

Take care not to exchange GND and 5V stick of Neo-Pixel LED strip while interfacing it to Arduino, else it will harm the NeoPixel LED strip. Likewise note that here we have utilized Arduino Mega however you can utilize some other Arduino model.

Associations with Arduino to NeoPixel RGB LED

Strip:

Arduino Pins	NeoPixel Strip's Pins
5v	5v
GND	GND
Digital Pin no. 36	DI (data in)

Working Explanation:

Working of NeoPixel LED with Arduino is simple. Simply tap the touch catch of any shading on the LCD in which you need to sparkle the NeoPixel LED strip. The LED will light as indicated by that shading. Code is written in such a manner in this way, that you can rehash this assignment unendingly without reseting the Arduino Mega. You can see the Code toward the finish of this article.

At the point when any catch is tapped on the LCD, information is sent to Arduino as well as Arduino further sends guidance to NeoPixel Strip to light likewise. For instance NeoPixel LED strip sparkles in

Green shading when we tap the Green catch on the LCD and LED strip gleams in Red shading when we press the Red catch, etc.

Programming Explanation:

To Interface TFT LCD with Arduino we have utilized a few libraries. Every one of the libraries come in one rar record and can be downloaded from this connection. Snap on 'Clone or download' and 'Download ZIP' record and add to your Arduino library organizer. This library is required for appropriate working of TFT LCD.

```
#include <SPFD5408_Adafruit_GFX.h>     // Core graphics library

#include  <SPFD5408_Adafruit_TFTLCD.h>   // Hardware-specific library

#include <SPFD5408_TouchScreen.h>
```

You should test your TFT LCD by copying Arduino with models codes given in the Library and check if codes are working appropriately. First check the illustrations test, at that point align test lastly paint test. In the event that you locate that all highlights all working fine, at that point start with code given in this instructional exercise.

Likewise for appropriate working of NeoPixel RGB LED strip, you will require one more library, which can be downloaded from here.

```
#include <Adafruit_NeoPixel.h>
```

As portrayed before Digital Pin 36 of MEGA is associated with DI stick of NeoPixel LED Strip as appeared in code underneath. Likewise the quantities of LEDs in the Strip are 30 so as is written in code:

```
#define PIN 36

#define NUM_LEDS 30
```

Show shades of LCD catches are meant by certain Codes. You can change these codes as indicated by your LCD.

```
#define BLACK  0x0000

#define YELLOW  0x001F

#define GREEN   0xF800

#define RED  0x07E0
```

```
#define CYAN   0x07FF

#define MAGENTA 0xF81F

#define BLUE  0xFFE0

#define WHITE  0xFFFF
```

A few parameters for the catches like size and position are characterized in the code:

```
uint16_t width = 0;

uint16_t height = 0;

uint16_t x = 40;

uint16_t y = height - 20;

uint16_t w = 75;

uint16_t h = 20;
```

h parameter is utilized to alter the size of catch on LCD. In the event that you make it 40, at that point the size of catch will get multiplied. y parameter is y organize of LCD.

Contact catches are signified by numbers as appeared in code:

```
#define BUTTONS 9

#define BUTTON_Red 0

#define BUTTON_DarkRed 1

#define BUTTON_RED 2

#define BUTTON_DarkGreen 3

#define BUTTON_DeepRed 4

#define BUTTON_Blue 5

#define BUTTON_LightBlue 6

#define BUTTON_LightBlue1 7
```

A few capacities are utilized to produce the shading out of NeoPixel like:

```
void EmitCyan();

void EmitWhite();

void EmitGreen();
```

```
void EmitYellow();

void EmitPink();

void EmitBlack();
```

To see the computerized RGB values as entered for the given shading, you can pursue this connection. Simply enter the shading you need your NeoPixel strip to sparkle, discover the RGB esteems for that shading and put in above capacities.

void initializeButtons() work is utilized for giving content and shading to catches and furthermore for setting them at required place on LCD.

```
void initializeButtons() {

  uint16_t x = 40;

  uint16_t y = height - 20;

  uint16_t w = 75;

  uint16_t h = 40;

  uint8_t spacing_x = 5

  ..... .....
```

....

void showCalibration() work is utilized to draw the catches on the LCD.

```
void showCalibration() {

  tft.setCursor (40, 0);

    for (uint8_t i = 0; i < 8; i++) {

      buttons[i].drawButton();

  }

}
```

For Glowing the NeoPixel LED strip in wanted Color the full Arduino code is given underneath. Code is bit protracted yet basic, you can comprehend the code effectively.

Code

```
#include <Adafruit_NeoPixel.h>
#include <math.h>
#include <SPFD5408_Adafruit_GFX.h>       // Core graphics library
#include <SPFD5408_Adafruit_TFTLCD.h> // Hardware-specific library
```

```
#include <SPFD5408_TouchScreen.h>
// *** SPFD5408 change -- End
#if defined(__SAM3X8E__)
#undef __FlashStringHelper::F(string_literal)
#define F(string_literal) string_literal
#endif
#define YP A1  // must be an analog pin, use "An" notation!
#define XM A2  // must be an analog pin, use "An" notation!
#define YM 7   // can be a digital pin
#define XP 6   // can be a digital pin
#define PIN 36
#define NUM_LEDS 30
#define BRIGHTNESS 80

// Calibrate values
#define TS_MINX 125
#define TS_MINY 85
#define TS_MAXX 965
#define TS_MAXY 905

// For better pressure precision, we need to know the resistance
// between X+ and X- Use any multimeter to read it
// For the one we're using, its 300 ohms across the X plate
TouchScreen ts = TouchScreen(XP, YP, XM, YM, 300);
#define LCD_CS A3
#define LCD_CD A2
#define LCD_WR A1
#define LCD_RD A0
```

```
// optional
#define LCD_RESET A4
// Assign human-readable names to some common
16-bit color values:
#define BLACK  0x0000
#define YELLOW  0x001F
#define GREEN  0xF800
#define RED  0x07E0
#define CYAN  0x07FF
#define MAGENTA 0xF81F
#define BLUE 0xFFE0
#define WHITE  0xFFFF

void EmitCyan();
void EmitWhite();
void EmitGreen();
void EmitYellow();
void EmitPink();
void EmitBlack();

uint16_t width = 0;
uint16_t height = 0;
uint16_t x = 40;
uint16_t y = height - 20;
uint16_t w = 75;
uint16_t h = 20;

Adafruit_TFTLCD tft(LCD_CS, LCD_CD, LCD_WR,
LCD_RD, LCD_RESET);
Adafruit_NeoPixel strip = Adafruit_NeoPixel-
(NUM_LEDS, PIN, NEO_GRB + NEO_KHZ800);
#define BOXSIZE 40
#define PENRADIUS 3
```

```
#define BUTTONS 9
#define BUTTON_Red 0
#define BUTTON_DarkRed 1
#define BUTTON_RED 2
#define BUTTON_DarkGreen 3
#define BUTTON_DeepRed 4
#define BUTTON_Blue 5
#define BUTTON_LightBlue 6
#define BUTTON_LightBlue1 7

Adafruit_GFX_Button buttons[BUTTONS];

uint16_t buttons_y = 0;
#define MINPRESSURE 10
#define MAXPRESSURE 1000

void setup() {
 // put your setup code here, to run once:
 Serial.begin(9600);
 Serial.println(F("Paint!"));
 tft.reset();
 tft.begin(0x9341); // SDFP5408
 strip.begin();
 strip.show(); // Initialize all pixels to 'off'
  tft.setRotation(0); // Need for the Mega, please
changed for your choice or rotation initial
 tft.fillScreen(BLACK);
 tft.setCursor (40, 20);
 tft.setTextSize (5);
 tft.setTextColor(WHITE);
 tft.println("COLORS");
 tft.setCursor (65, 85);
 width = tft.width() - 1;
```

```
height = tft.height() - 100;
initializeButtons();
showCalibration();
}
void loop() {
 // put your main code here, to run repeatedly:
 // Test of calibration

  int i = 0;
 TSPoint p;
 // Wait a touch
 digitalWrite(13, HIGH);
 p = waitOneTouch();
 digitalWrite(13, LOW);
 p.x = mapXValue(p);
 p.y = mapYValue(p);
 for (uint8_t b = 0; b < BUTTONS; b++) {
  if (buttons[b].contains(p.x, p.y)) {
  // Action
  switch (b) {
  case BUTTON_Red:
   EmitPink();
   showCalibration();
  break;
  case BUTTON_DarkRed:
   EmitCyan();
   showCalibration();
  break;
```

```
    case BUTTON_RED:
    EmitBlack();
    showCalibration();
   break;


    case BUTTON_DarkGreen:
    EmitGreen();
    showCalibration();
   break;


    case BUTTON_Blue:
    EmitBlue();
    showCalibration();
   break;


    case BUTTON_LightBlue:
    EmitYellow();
    showCalibration();
   break;


    case BUTTON_DeepRed:
    EmitDeepRed();
    showCalibration();
   break;
  }
 }
}
```

```
}
void initializeButtons() {
 uint16_t x = 40;
 uint16_t y = height - 20;
 uint16_t w = 75;
 uint16_t h = 40;
 uint8_t spacing_x = 5;
 uint8_t textSize = 2;
 char buttonlabels[7][20] = {"PINK", "CYAN", "WHITE",
"GREEN", "RED", "BLUE", "YELLOW"};
 uint16_t buttoncolors[15] = {RED, GREEN, BLACK,
MAGENTA, CYAN, BLUE, YELLOW};
 for (uint8_t b = 0; b < 9; b++) {
  if (b < 3)
  {
   buttons[b].initButton(&tft,            // TFT object
          x + b * (w + spacing_x), y,     // x, y,
            w, h, BLACK, buttoncolors[b], WHITE,   // w,
h, outline, fill,
            buttonlabels[b], textSize);
   Serial.print( h);
  } // text
  if (b == 3)
  {
   uint16_t x = 40;
   uint16_t y = height + 30 ;
   uint16_t w = 75;
   uint16_t h = 40;
   buttons[b].initButton(&tft,            // TFT object
          x + 0 * (w + spacing_x), y,     // x, y,
```

```
                w, h, BLACK, buttoncolors[b], WHITE,   // w,
h, outline, fill,
                buttonlabels[b], textSize);
   }
   if (b == 4)
   {
     uint16_t x = 40;
     uint16_t y = height + 30 ;
     uint16_t w = 75;
     uint16_t h = 40;
     buttons[b].initButton(&tft,              // TFT object
            x + 1 * (w + spacing_x) , y,    // x, y,
            w, h, BLACK, buttoncolors[b], WHITE,   // w,
h, outline, fill,
                buttonlabels[b], textSize);
   }
   if (b == 5)
   {
     uint16_t x = 40;
     uint16_t y = height + 30 ;
     uint16_t w = 75;
     uint16_t h = 40;
     buttons[b].initButton(&tft,              // TFT object
            x + 2 * (w + spacing_x) , y,    // x, y,
            w, h, BLACK, buttoncolors[b], WHITE,   // w,
h, outline, fill,
                buttonlabels[b], textSize);
   }
   if (b == 6)
   {
```

```
  uint16_t x = 40;
  uint16_t y = height + 80 ;
  uint16_t w = 75;
  uint16_t h = 20;
  buttons[b].initButton(&tft,               // TFT object
          x + 0 * (w + spacing_x) , y,     // x, y,
          w, h,BLACK, buttoncolors[b], WHITE,   // w,
h, outline, fill,
          buttonlabels[b], textSize);
  }
  if(b == 7)
  {
  uint16_t x = 40;
  uint16_t y = height + 80 ;
  uint16_t w = 75;
  uint16_t h = 40;
  buttons[b].initButton(&tft,               // TFT object
          x + 2 * (w + spacing_x) , y,     // x, y,
          w, h,BLACK, buttoncolors[b], WHITE,   // w,
h, outline, fill,
          buttonlabels[b], textSize);
  }
  }
  // Save the y position to avoid draws
  buttons_y = y;
}
// Map the coordinate X
uint16_t mapXValue(TSPoint p) {
  uint16_t x = map(p.x, TS_MINX, TS_MAXX, 0, tft.
width());
```

```
//Correct offset of touch. Manual calibration
//x+=1;
return x;
}
uint16_t mapYValue(TSPoint p) {
    uint16_t y = map(p.y, TS_MINY, TS_MAXY, 0,
tft.height());
//Correct offset of touch. Manual calibration
//y-=2;
return y;
}
TSPoint waitOneTouch() {
TSPoint p;
do {
  p = ts.getPoint();
  pinMode(XM, OUTPUT); //Pins configures again for
TFT control
  pinMode(YP, OUTPUT);
  } while ((p.z < MINPRESSURE ) || (p.z > MAXPRES-
SURE));
return p;
}
void showCalibration() {
tft.setCursor (40, 0);
  for (uint8_t i = 0; i < 8; i++) {
    buttons[i].drawButton();
  }
}
void EmitDeepRed()
{
```

```
for(int i=0;i<30;i++)
{
strip. setPixelColor(i, 255, 0, 0);
strip.show();
}
}
void EmitCyan()
{
 for(int i=0;i<30;i++)
 {
 strip. setPixelColor(i,0, 255, 255);
 strip.show();
 }
}
void EmitWhite()
{
 for(int i=0;i<30;i++)
 {
 strip. setPixelColor(i,255, 255, 255);
 strip.show();
 }
}
void EmitGreen()
{
 for(int i=0;i<30;i++)
 {
 strip. setPixelColor(i,0, 255, 0);
 strip.show();
 }
}
```

```
void EmitBlue()
{
 for(int i=0;i<30;i++)
 {
 strip. setPixelColor(i,0, 0, 255);
 strip.show();
 }
}
void EmitYellow()
{
 for(int i=0;i<30;i++)
 {
 strip. setPixelColor(i,255, 255, 0);
 strip.show();
 }
}
void EmitPink()
{
 for(int i=0;i<30;i++)
 {
 strip. setPixelColor(i,255, 0, 255);
 strip.show();
 }
}
void EmitBlack()
{
 for(int i=0;i<30;i++)
 {
 strip. setPixelColor(i,255, 255, 255);
 strip.show();
```

```
  }
}
```

◆ ◆ ◆

7. PING PONG GAME UTILIZING ARDUINO AND ACCELEROMETER

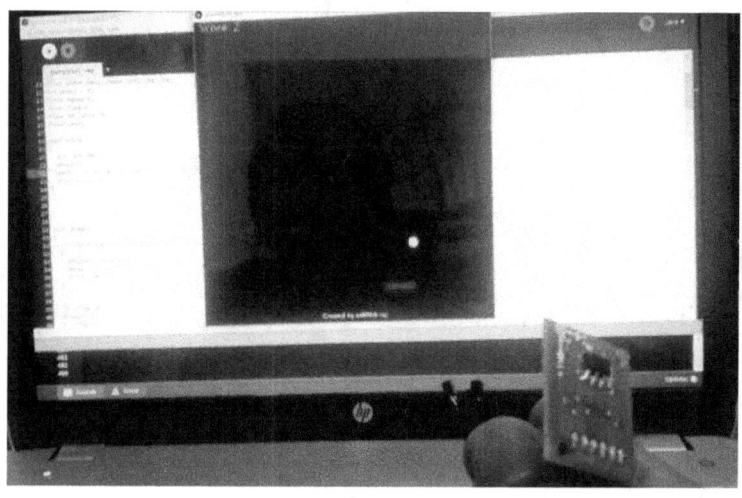

Increased Reality and Virtual Gaming has become an ongoing pattern in the gaming business. The hours of utilizing a console/Joystick and a mouse to play a PC game has gone behind. Presently every gaming console accompanies a Virtual Controller that encourages us to play the game utilizing our body developments and motions, along these lines the gaming experience has expanded a great deal and client feels increasingly included into the game.

In this undertaking we should attempt to have a great time as we learn through the task. Give us a chance to make a game (Yes you heard me right we are goanna make a game) and play it utilizing your hand's development. We are making the great Ping Pong Ball Game utilizing Arduino and Accelerometer.

Overview:

There are huge amounts of open source programming's accessible nowadays which has brought heaps of joy for specialists like us, and Processing is one of them. With this JAVA based application we can work out possess programming (.exe position) and furthermore an android application (.apk record). So we are going to utilize this product to assemble our game, we have recently utilized Processing in making Arduino Chat Room.

The equipment part will comprise an Arduino which will bring the contribution from an Accelerometer to bolster it sequentially to our PC/Laptop.

So how about we go shopping!!!!

Components Required:

- Arduino (any variant or modcl)
- Connecting wires
- Accelerometer
- Interest (Lolz)

Accelerometer and Arduino Nano

Circuit Explanation:

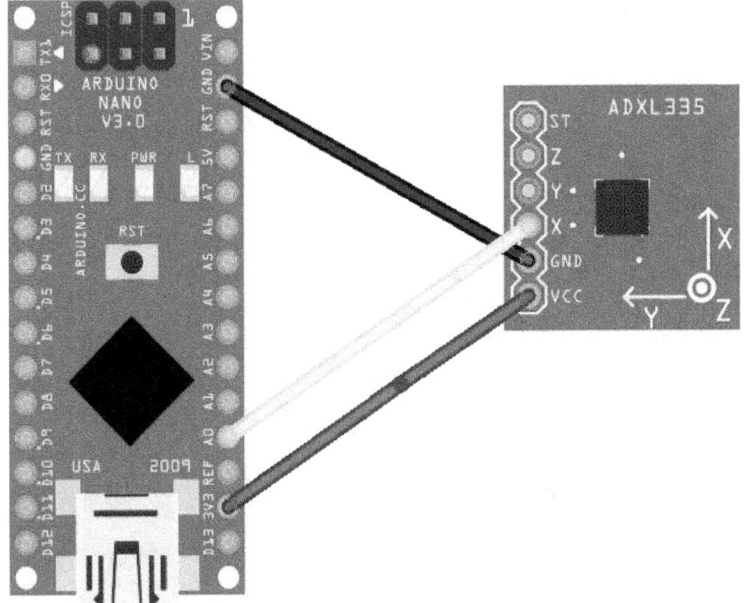

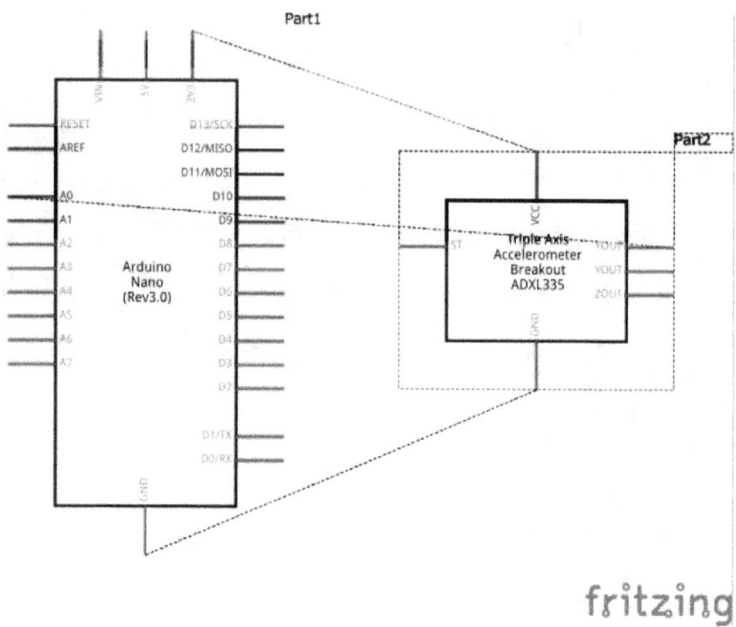

Circuit of Arduino Ping Pong Ball Game Project doesn't include any perplexing associations. I have utilized an Arduino Nano with an Accelerometer. In any case, there is scarcely any things to be dealt with as referenced beneath:

1. Your Accelerometer can't deal with 5V, so consistently associate the Vcc of accelerometer to your 3.3V stick of Arduino.

2. Each Accelerometer experiences the impact of gravity which must be taken care of while programming (basically utilizing a channel).

In light of this current we should investigate the working of an Accelerometer and how we use it.

Working of Accelerometer:

An Accelerometer is a gadget which can change over quickening toward any path to its particular variable voltage. This is cultivated by utilizing capacitors (allude picture), as the Accel moves, the capacitor present inside it, will likewise experience changes (allude picture) in view of the development, since the capacitance is shifted, a variable voltage can likewise be acquired.

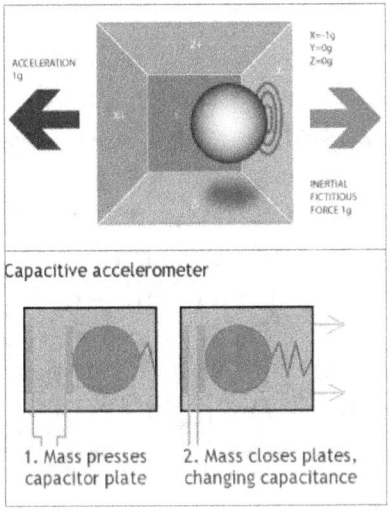

Along these lines, as referenced over each accelerometer experiences the issue of gravity impact. Regardless of how precise your sensor is adjusted (even your apple telephones Accel.), it will be influenced by gravity. An increasingly specialized clarification for this issue is given beneath.

"Theoretically, an increasing speed sensor decides the quickening that is applied to a gadget (Ad) by estimating the powers that are applied to the sensor itself (Fs) utilizing the accompanying relationship:

Advertisement = - ?Fs/mass

Be that as it may, the power of gravity is continually impacting the deliberate quickening as indicated by the accompanying relationship:

Advertisement = - g - ?F/mass

Hence, when the gadget is perched on a table (and not quickening), the accelerometer peruses a size of g = 9.81 m/s2. Additionally, when the gadget is in free fall and consequently quickly quickening toward the ground at 9.81 m/s2, its accelerometer peruses a greatness of g = 0 m/s2. Along these lines, to gauge the genuine increasing speed of the gadget, the commitment of the power of gravity must be expelled from the accelerometer information. This can be accomplished by applying a high-pass channel. At the mean time, a low-pass channel can be utilized to disengage the power of gravity."

Source: Android designers

Presently, in Arduino we can diminish the impact of gravity by utilizing a Simple Filter. This channel will comprise of two exhibits, one is utilized to store the example esteems from sensor and the other is utilized to sort the example esteems, and locate the most rehashed worth. Give us a chance to execute this calculation in our Arduino and prepare our equipment.

Programming Arduino:

The Arduino program is given beneath in Code area. There is no basic information that must be adjusted. Be that as it may, you should think about the accompanying:

Increment the example size if your Accel still shows irregular qualities.

```
#define Samplesize  13      // filterSample number
```

Play with the 9600 baud rate to build the speed of correspondence among Arduino and Processing. However, ensure you change them in both the product (Programs).

```
void setup(){

  Serial.begin(9600);

}
```

My Accelerometer on X-hub gives 193 on far left end and 280 on far right end, measure them for your Accel and update the worth.

> toSend = map (smoothData1, 193, 280, 0, 255);

The qualities are mapped into a solitary byte of information for sequential correspondence.

Further check the Comments in the beneath offered Code to comprehend it unmistakably.

Programming Processing:

Preparing is open source programming which is utilized by specialists for Graphics planning. This product is utilized to create programming and Android applications. It is very simple to create and particularly like the Android Development IDE. Consequently I have abbreviated the clarification.

The Processing Code for the Ping Pong Game is given here:

- Handling Code for Arduino Ping Ball Game

Right click on it and snap on 'Spare connect as..' to download the code document. At that point open the record in 'Handling' programming and snap on 'Run' catch to play the Game. You have to introduce 'Preparing' programming to open *.pde records.

Underneath line, in the void arrangement() capacity of Processing code is significant, as it chooses from which port to information from.

```
port = new Serial(this,Serial.list()[4],9600);    //
Reads the 4th PORT at 9600 baudrate
```

Here I have perused information from the fourth port from my Arduino.

So for Example on the off chance that you have COM[5] COM[2] COM[1] COM [7] COM[19]

At that point the above code will peruse information from COM[7].

Testing:

Presently since our Processing and Arduino sketch is prepared, simply transfer the underneath offered program to Arduino and associate your Arduino to client PC careful programming link and dispatch the game by Run the Processing code record (.pde). That is it! Move your Accelerometer and play your Ping Pong Game.

When you have comprehended the program you can make numerous comparable games and play them utilizing your Arduino, Further the Y-pivot and Z-hub may likewise be incorporated for gaming.

Code

```
/*
 * Program for filtering the X-axis values of accel and transmitting it serially

 */
#define AccelPin   A0     // A0 is connected to X-axis of Accel
#define Samplesize  13     // filterSample number
int Array1 [Samplesize];        // array for holding raw sensor values for sensor
int rawData1, smoothData1;    // variables for sensor data
int toSend;
void setup(){
 Serial.begin(9600);
```

```
}
void loop()
{
 rawData1 = analogRead(AccelPin);            // read
X-axis of accelerometer
 smoothData1 = digitalSmooth(rawData1, Array1);
 toSend = map (smoothData1, 193, 280, 0, 255);      //
the data from accelerometer mapped to form a byte
 Serial.write (toSend);
 delay(100);
}
 int digitalSmooth(int rawIn, int *sensSmoothArray)
{     // "int *sensSmoothArray" passes an array to the
function - the asterisk indicates the array name is a
pointer
 int j, k, temp, top, bottom;
 long total;
 static int i;
 static int sorted[Samplesize];
 boolean done;
 i = (i + 1) % Samplesize;          // increment counter
and roll over if necc. - % (modulo operator) rolls over
variable
 sensSmoothArray[i] = rawIn;        // input new data
into the oldest slot
 for (j=0; j<Samplesize; j++){      // transfer data array
into anther array for sorting and averaging
  sorted[j] = sensSmoothArray[j];
 }
 done = 0;                // flag to know when we're done
```

sorting

```
while(done != 1){      // simple swap sort, sorts num-
bers from lowest to highest
 done = 1;
 for (j = 0; j < (Samplesize - 1); j++){
  if (sorted[j] > sorted[j + 1]){      // numbers are out of
order - swap
   temp = sorted[j + 1];
   sorted [j+1] = sorted[j];
   sorted [j] = temp;
   done = 0;
  }
 }
}
bottom = max((((Samplesize * 15) / 100), 1);
top = min(((((Samplesize * 85)/100) + 1 ), (Samplesize
- 1));  // the + 1 is to make up for asymmetry caused by
integer rounding
 k = 0;
 total = 0;
 for ( j = bottom; j< top; j++){
  total += sorted[j];     // total remaining indices
  k++;
 }
 return total / k;      // divide by number of samples
}
```

◆ ◆ ◆

8. CALL AND MESSAGE UTILIZING ARDUINO AS WELL AS GSM MODULE

Some of the time individuals think that its hard to utilize the GSM Module for its essential capacities like calling, messaging and so forth., explicitly with the Microcontrollers. So here we are going to manufacture a Simple Mobile Phone utilizing Arduino, in which GSM Module is utilized to Make the Call, answer the Call, send SMS, and read SMS, and furthermore this Arduino telephone has Mic as well as Speaker to talk over this Phone. This venture will likewise fill in as a legitimate interfacing of GSM Module with Arduino, with all the Code expected to work any Phone's essential capacities.

Components Required:

- Arduino Uno
- GSM Module SIM900
- 16x2 LCD
- Power supply
- 4x4 Keypad
- Connecting jumper wire
- Breadboard or PCB
- Speaker
- MIC
- SIM Card

Working Explanation:

In this Arduino Mobile Phone Project, we have utilized Arduino Uno to control entire framework's highlights and interfacing every one of the parts in this framework. A 4x4 Alphanumeric Keypad is utilized for taking all sort of data sources like: Enter portable number, type messages, cause a call, to get a call, send SMS, read SMS and so forth. GSM Module is utilized to speak with the system for calling and informing reason. We have additionally interfaced a MIC and a Speaker for Voice Call and Ring sound and a 16x2 LCD is utilized for indicating messages, guidelines and cautions.

Alphanumeric is a strategy to enter numbers and letters in order both by utilizing same keypad. In this technique, we have interfaced 4x4 keypad with Arduino and composed Code for tolerating letter sets as well, check the Code in Code area underneath.

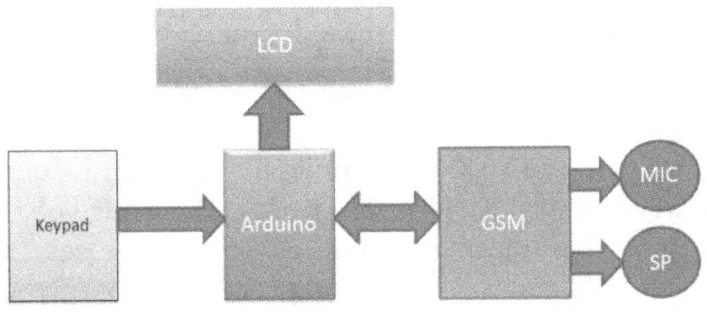

Working of this undertaking is simple. Every one of the highlights will be performed by Using Alphanumeric Keypad. Check the Full code beneath to appropriately comprehend the procedure. Here we will clarify all the four highlights of the tasks beneath.

Explaining Four Features of Arduino Mobile Phone:

1. Make a Call:

To make a call by utilizing our Arduino based Phone, we need to squeeze 'C' and afterward need to enter the Mobile Number on which we need to make a call. Number will be entered by utilizing alphanumeric keypad. In the wake of entering the number we again need to squeeze 'C'. Presently Arduino will process for interfacing the call to the entered number by utilizing AT direction:

```
ATDxxxxxxxxxx; <Enter>    where xxxxxxxxx is
entered Mobile Number.
```

2. Get a Call:

Getting a call is extremely simple. At the point when somebody is calling to your framework SIM number, which is there in GSM Module, at that point your framework will show 'Approaching... ' message over the LCD with approaching number of guest. Presently we simply need to Press 'A' to go to this call. At the point when we press 'An', Arduino will send offered order to GSM Module:

```
ATA <enter>
```

3. Send SMS:

Exactly when we need to send a SMS utilizing our Arduino based Mobile Phone, at that point we have to Press 'B'. Presently System will request Recipient Number, signifies 'to whom' we need to send SMS. As a result of entering the number we have to squeeze 'D' and now LCD requests message. Presently we have to type the message, similar to we enter in typical portable, by utilizing keypad and afterward in the wake of entering the message we have to squeeze 'D' to send SMS. To Send SMS Arduino sends given order:

AT+CMGF=1 <enter>

AT+CMGS="xxxxxxxxxx" <enter>   where: xxxxx-xxxxx is entered mobile number

Also, send 26 to GSM to send SMS.

4. Get and Read SMS:

This element is likewise basic. In this, GSM will get SMS and stores it in SIM card. What's more, Arduino constantly screens the got SMS sign over UART. We simply need to Press 'D', to peruse the SMS, when we see the New Message image on the LCD. The following is the SMS Received sign shown on the Serial port is:

+CMTI: "SM" <SMS stored location>

+CMTI: "SM",6 Where 6 is message location where it stored in SIM card.

When Arduino gets this 'SMS got' sign then it extricates SMS putting away area and sends direction to GSM to peruse the got SMS. What's more, show 'Another Message Symbol' over the LCD.

AT+CMGR=<SMS stored location><enter>

AT+CMGR=6

Presently GSM sends put away message to Arduino and afterward Arduino extricate principle SMS and show it over the LCD and afterward subsequent to perusing this SMS Arduino Clear the 'New SMS image' from the LCD.

Note: There is no coding for MIC as well as Speaker.

Check the Full code beneath to appropriately comprehend the procedure.

Circuit Diagram and Explanation:

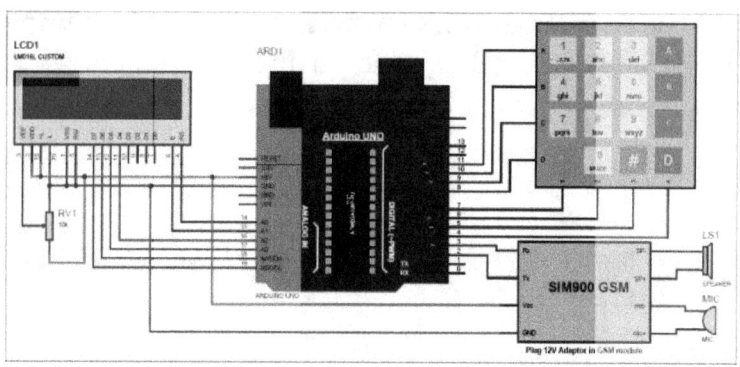

Circuit Diagram of this for interfacing GSM SIM900 and Arduino is given previously. 16x2 LCD pins RS, EN, D4, D5, D6 and D7 are associated with stick number 14, 15, 16, 17, 18 and 19 of Arduino separately. GSM Module's Rx and Tx pins are legitimately associated with Arduino's stick D3 and D2 individually (Ground of Arduino and GSM must be associated with one another). 4x4 keypad Row pins R1, R2, R3, R4 are straightforwardly connected to stick number 11,10, 9, 8 of Arduino and Colum pins of keypad C1, C2, C3 are connected with stick number 7, 6, 5, 4 of Arduino. MIC is straightforwardly associated at mic+ and mic- of GSM Module and Speaker is legitimately associated at SP+ and SP-pins for GSM Module.

Programming Explanation:

Programming some portion of this undertaking is minimal complex for tenderfoots. In this code we have utilized keypad library #include <Keypad.h> for interfacing basic keypad for entering numbers. What's more, for entering letter sets with a similar keypad, we have made capacity void alfakey(). We have made each key multi working as well as we can enter any character or number by utilizing just 10 keys.

Like in case we press key 2 (abc2), it will show 'an' and in the event that we squeezes it once more, at that point it will supplant 'a' to 'b' and in case again we press multiple times, at that point it will show 'c' at same place in LCD. In case we sit tight for some time subsequent to squeezing key, cursor will programmed move to next position in LCD. Presently we can enter next scorch or number. A similar methodology is applied for different keys.

```
#include <Keypad.h>

const byte ROWS = 4; //four rows

const byte COLS = 4; //four columns

char hexaKeys[ROWS][COLS] =

{

 {'1','2','3','A'},

 {'4','5','6','B'},

 {'7','8','9','C'},

 {'*','0','#','D'}

};

byte rowPins[ROWS] = {11, 10, 9, 8}; //connect to
the row pinouts of the keypad

byte colPins[COLS] = {7, 6, 5, 4}; //connect to the
column pinouts of the keypad
```

```
//initialize an instance of class NewKeypad

Keypad            customKeypad            =
Keypad( makeKeymap(hexaKeys), rowPins, col-
Pins, ROWS, COLS);
```

```
void alfakey()

{

int x=0,y=0;

int num=0;

 while(1)

 {

  lcd.cursor();

  char key=customKeypad.getKey();

  if(key)

  {

   if(key=='1')
```

```
    {

    num=0;

    lcd.setCursor(x,y);

    .... .....

    ........ ....
```

Aside from working keypad, we have made numerous different capacities like void call() for calling highlight of Phone, void sms() for informing highlight, void lcd_status() for show LCD status void gsm_init() for introducing the GSM Module and so on. Check underneath all other capacity identified with make and get Call and send and read SMS utilizing GSM Module and Arduino. Every one of the capacities are simple and reasonable.

Code

```
#include <SoftwareSerial.h>
SoftwareSerial Serial1(2, 3); // RX, TX
#include<LiquidCrystal.h>
LiquidCrystal lcd(14,15,16,17,18,19);
byte back[8] =
{
 0b00000,
 0b00000,
 0b11111,
```

```
0b10101,
0b11011,
0b11111,
0b00000,
0b00000
};
String number="";
String msg="";
String instr="";
String str_sms="";
String str1="";
int ring=0;
int i=0,temp=0;
int sms_flag=0;
char sms_num[3];
int rec_read=0;
int temp1=0;
#include <Keypad.h>
const byte ROWS = 4; //four rows
const byte COLS = 4; //four columns
char hexaKeys[ROWS][COLS] =
{
 {'1','2','3','A'},
 {'4','5','6','B'},
 {'7','8','9','C'},
 {'*','0','#','D'}
};
byte rowPins[ROWS] = {11, 10, 9, 8}; //connect to the
row pinouts of the keypad
byte colPins[COLS] = {7, 6, 5, 4}; //connect to the col-
```

umn pinouts of the keypad

```
//initialize an instance of class NewKeypad
Keypad           customKeypad           =
Keypad( makeKeymap(hexaKeys), rowPins, colPins,
ROWS, COLS);
String    ch="1,.?!@abc2def3ghi4jkl5mno6pqrs7tu-
v8wxyz90";
void setup()
{
 Serial1.begin(9600);
 lcd.begin(16,2);
 lcd.createChar(1, back);
 lcd.print("Simple Mobile ");
 lcd.setCursor(0,1);
 lcd.print("System Ready..");
 delay(1000);
 gsm_init();
 lcd.clear();
 lcd.print("System Ready");
 delay(2000);
}
void loop()
{
 serialEvent();
 if(sms_flag==1)
 {
  lcd.clear();
  lcd.print("New Message");
  int ind=instr.indexOf("+CMTI: \"SM\",");
  ind+=12;
```

```
int k=0;
lcd.setCursor(0,1);
lcd.print(ind);
while(1)
{
 while(instr[ind]!= 0x0D)
 {
  sms_num[k++]=instr[ind++];
 }
 break;
}
ind=0;
sms_flag=0;
lcd.setCursor(0,1);
lcd.print("Read SMS --> D");
delay(4000);
instr="";
rec_read=1;
temp1=1;
i=0;
}


if(ring == 1)
{
number="";
int loc=instr.indexOf("+CLIP: \"");
if(loc > 0)
{
 number+=instr.substring(loc+8,loc+13+8);
}
```

```
 lcd.setCursor(0,0);
 lcd.print("Incomming...  ");
 lcd.setCursor(0,1);
 lcd.print(number);
 instr="";
 i=0;
}
else
{
serialEvent();
lcd.setCursor(0,0);
lcd.print("Call --> C   ");
lcd.setCursor(0,1);
lcd.print("SMS --> B ");
if(rec_read==1)
{
 lcd.write(1);
 lcd.print("  ");
}
else
lcd.print("   ");
}


  char key=customKeypad.getKey();
if(key)
{
 if(key== 'A')
 {
  if(ring==1)
  {
```

```
 Serial1.println("ATA");
 delay(5000);
 }
}
else if(key=='C')
{
 call();
}
else if(key=='B')
{
 sms();
}
else if(key == 'D' && temp1==1)
{
 rec_read=0;
 lcd.clear();
 lcd.print("Please wait...");
 Serial1.print("AT+CMGR=");
 Serial1.println(sms_num);
 int sms_read_flag=1;
 str_sms="";
 while(sms_read_flag)
 {
  while(Serial1.available()>0)
  {
   char ch=Serial1.read();
   str_sms+=ch;
   if(str_sms.indexOf("OK")>0)
   {
    sms_read_flag=0;
```

```
      //break;
    }
   }
  }
  int l1=str_sms.indexOf("\"\r\n");
  int l2=str_sms.indexOf("OK");
  String sms=str_sms.substring(l1+3,l2-4);
  lcd.clear();
  lcd.print(sms);
  delay(5000);
  }
  delay(1000);
 }
}
void call()
{
 number="";
 lcd.clear();
 lcd.print("After Enter No.");
 lcd.setCursor(0,1);
 lcd.print("Press C to Call");
 delay(2000);
 lcd.clear();
 lcd.print("Enter Number:");
 lcd.setCursor(0,1);
 while(1)
 {
  serialEvent();
  char key=customKeypad.getKey();
  if(key)
```

```
{
 if(key=='C')
 {
  lcd.clear();
  lcd.print("Calling........");
  lcd.setCursor(0,1);
  lcd.print(number);
  Serial1.print("ATD");
  Serial1.print(number);
  Serial1.println(";");
  long stime=millis()+5000;
  int ans=1;
  while(ans==1)
  {
   while(Serial1.available()>0)
   {
    if(Serial1.find("OK"))
    {
     lcd.clear();
     lcd.print("Ringing....");
     int l=0;
     str1="";
     while(ans==1)
     {
      while(Serial1.available()>0)
      {
       char ch=Serial1.read();
       str1+=ch;
       if(str1.indexOf("NO CARRIER")>0)
       {
```

```
        lcd.clear();
        lcd.print("Call End");
        delay(2000);
        ans=0;
        return;
        }
        }
        char key=customKeypad.getKey();
        if(key == 'D')
        {
        lcd.clear();
        lcd.print("Call End");
        delay(2000);
        ans=0;
        return;
        }
        if(ans==0)
        break;
        }
        }
        }
    }
    }
    else
    {
    number+=key;
    lcd.print(key);
    }
    }
}
```

```
}
void sms()
{
 lcd.clear();
 lcd.print("Initilising SMS");
 Serial1.println("AT+CMGF=1");
 delay(400);
 lcd.clear();
 lcd.print("After Enter No.");
 lcd.setCursor(0,1);
 lcd.print("Press D     ");
 delay(2000);
 lcd.clear();
 lcd.print("Enter Rcpt No.:");
 lcd.setCursor(0,1);
 Serial1.print("AT+CMGS=\"");
 while(1)
 {
  serialEvent();
  char key=customKeypad.getKey();
  if(key)
  {
   if(key=='D')
   {
    //number+="";
    Serial1.println("\"");
    break;
   }
   else
   {
```

```
  //number+=key;
  Serial1.print(key);
  lcd.print(key);
 }
}
}
lcd.clear();
lcd.print("After Enter MSG ");
lcd.setCursor(0,1);
lcd.print("Press D to Send ");
delay(2000);
lcd.clear();
lcd.print("Enter Your Msg");
delay(1000);
lcd.clear();
lcd.setCursor(0,0);
alfakey();
}
void alfakey()
{
int x=0,y=0;
int num=0;
while(1)
{
 lcd.cursor();
 char key=customKeypad.getKey();
 if(key)
 {
  if(key=='1')
  {
```

```
num=0;
lcd.setCursor(x,y);
lcd.print(ch[num]);
for(int i=0;i<3000;i++)
{
lcd.noCursor();
char key=customKeypad.getKey();
if(key=='1')
{
 num++;
 if(num>5)
 num=0;
 lcd.setCursor(x,y);
 lcd.print(ch[num]);
 i=0;
 delay(200);
}
}
x++;
if(x>15)
{
 x=0;
 y++;
 y%=2;
}
 msg+=ch[num];
}
else if(key=='2')
{
 num=6;
```

```
lcd.setCursor(x,y);
lcd.print(ch[num]);
for(int i=0;i<3000;i++)
{
lcd.noCursor();
char key=customKeypad.getKey();
if(key=='2')
{
num++;
if(num>9)
num=6;
lcd.setCursor(x,y);
lcd.print(ch[num]);
i=0;
delay(200);
}
}
x++;
if(x>15)
{
x=0;
y++;
y%=2;
}
msg+=ch[num];
}
else if(key=='3')
{
num=10;
lcd.setCursor(x,y);
```

```
lcd.print(ch[num]);
for(int i=0;i<3000;i++)
{
lcd.noCursor();
char key=customKeypad.getKey();
if(key=='3')
{
 num++;
 if(num>13)
 num=10;
 lcd.setCursor(x,y);
 lcd.print(ch[num]);
 i=0;
 delay(200);
}
}
x++;
 if(x>15)
{
 x=0;
 y++;
 y%=2;
}
 msg+=ch[num];
}
else if(key=='4')
{
 num=14;
 lcd.setCursor(x,y);
 lcd.print(ch[num]);
```

```
for(int i=0;i<3000;i++)
{
lcd.noCursor();
char key=customKeypad.getKey();
if(key=='4')
{
 num++;
 if(num>17)
 num=14;
 lcd.setCursor(x,y);
 lcd.print(ch[num]);
 i=0;
 delay(200);
 }
 }
 x++;
 if(x>15)
 {
  x=0;
  y++;
  y%=2;
 }
 msg+=ch[num];
 }
    else if(key=='5')
{
 num=18;
 lcd.setCursor(x,y);
 lcd.print(ch[num]);
 for(int i=0;i<3000;i++)
```

```
{
lcd.noCursor();
char key=customKeypad.getKey();
if(key=='5')
{
 num++;
 if(num>21)
 num=18;
 lcd.setCursor(x,y);
 lcd.print(ch[num]);
 i=0;
 delay(200);
 }
}
x++;
 if(x>15)
 {
 x=0;
 y++;
 y%=2;
 }
 msg+=ch[num];
 }
 else if(key=='6')
 {
 num=22;
 lcd.setCursor(x,y);
 lcd.print(ch[num]);
 for(int i=0;i<3000;i++)
 {
```

```
lcd.noCursor();
char key=customKeypad.getKey();
if(key=='6')
{
num++;
if(num>25)
num=22;
lcd.setCursor(x,y);
lcd.print(ch[num]);
i=0;
delay(200);
}
}
x++;
if(x>15)
{
x=0;
y++;
y%=2;
}
msg+=ch[num];
}
else if(key=='7')
{
num=26;
lcd.setCursor(x,y);
lcd.print(ch[num]);
for(int i=0;i<3000;i++)
{
lcd.noCursor();
```

```
char key=customKeypad.getKey();
if(key=='7')
{
num++;
if(num>30)
num=26;
lcd.setCursor(x,y);
lcd.print(ch[num]);
i=0;
delay(200);
}
}
x++;
if(x>15)
{
x=0;
y++;
y%=2;
}
msg+=ch[num];
}
else if(key=='8')
{
num=31;
lcd.setCursor(x,y);
lcd.print(ch[num]);
for(int i=0;i<3000;i++)
{
lcd.noCursor();
char key=customKeypad.getKey();
```

```
if(key=='8')
{
num++;
if(num>34)
num=31;
lcd.setCursor(x,y);
lcd.print(ch[num]);
i=0;
delay(200);
}
}
x++;
if(x>15)
{
x=0;
y++;
y%=2;
}
msg+=ch[num];
}
else if(key=='9')
{
num=35;
lcd.setCursor(x,y);
lcd.print(ch[num]);
for(int i=0;i<3000;i++)
{
lcd.noCursor();
char key=customKeypad.getKey();
if(key=='9')
```

```
{
num++;
if(num>39)
num=35;
lcd.setCursor(x,y);
lcd.print(ch[num]);
i=0;
delay(200);
}
}
x++;
if(x>15)
{
x=0;
y++;
y%=2;
}
msg+=ch[num];
}
else if(key=='0')
{
num=40;
lcd.setCursor(x,y);
lcd.print(ch[num]);
for(int i=0;i<3000;i++)
{
lcd.noCursor();
char key=customKeypad.getKey();
if(key=='0')
{
```

```
  num++;
  if(num>41)
  num=40;
  lcd.setCursor(x,y);
  lcd.print(ch[num]);
  i=0;
  delay(200);
  }
  }
 x++;
 if(x>15)
 {
  x=0;
  y++;
  y%=2;
 }
 msg+=ch[num];
 }
 else if(key=='D')
 {
  lcd.clear();
  lcd.print("Sending SMS....");
  // Serial1.print("AT+CMGS=");
  // Serial1.print(number);
  // delay(2000);
  Serial1.print(msg);
  Serial1.write(26);
  delay(5000);
  lcd.clear();
  lcd.print("SMS Sent to");
```

```
      lcd.setCursor(0,1);
      lcd.print(number);
      delay(2000);
      number="";
      break;
    }
   }
  }
}
void send_data(String message)
{
 Serial1.println(message);
 delay(200);
}
void send_sms()
{
 Serial1.write(26);
}
void lcd_status()
{
 lcd.setCursor(2,1);
 lcd.print("Message Sent");
 delay(2000);
 //lcd.setCursor()
 //lcd.print("")
 //return;
}
void back_button()
{
 //lcd.setCursor(0,15);
```

```
}
void ok_button()
{
 lcd.setCursor(0,4);
 lcd.print("OK");
}
void call_button()
{
 lcd.setCursor(0,4);
 lcd.print("CALL");
}
void sms_button()
{
 lcd.setCursor(0,13);
 lcd.print("SMS");
}
void gsm_init()
{
 lcd.clear();
 lcd.print("Finding Module..");
 boolean at_flag=1;
 while(at_flag)
 {
  Serial1.println("AT");
  while(Serial1.available()>0)
  {
   if(Serial1.find("OK"))
   at_flag=0;
  }
```

```
  delay(1000);
}
lcd.clear();
lcd.print("Module Connected..");
delay(1000);
lcd.clear();
lcd.print("Disabling ECHO");
boolean echo_flag=1;
while(echo_flag)
{
 Serial1.println("ATE1");
 while(Serial1.available()>0)
 {
  if(Serial1.find("OK"))
  echo_flag=0;
 }
 delay(1000);
}
lcd.clear();
lcd.print("Echo OFF");
delay(1000);
lcd.clear();
lcd.print("Finding Network..");
boolean net_flag=1;
while(net_flag)
{
 Serial1.println("AT+CPIN?");
 while(Serial1.available()>0)
 {
  if(Serial1.find("+CPIN: READY"))
```

```
   net_flag=0;
  }
  delay(1000);
 }
 lcd.clear();
 lcd.print("Network Found..");
 delay(1000);
 lcd.clear();
}
void serialEvent()
{
 while(Serial1.available())
 {
  char ch=Serial1.read();
  instr+=ch;
  i++;
  if(instr[i-4] == 'R' && instr[i-3] == 'T' && instr[i-2] ==
'N' && instr[i-1] == 'G' )
  {
   ring=1;
  }
  if(instr.indexOf("NO CARRIER")>=0)
  {
   ring=0;
   i=0;
  }
  if(instr.indexOf("+CMTI: \"SM\"")>=0)
  {
   sms_flag=1;
  }
```

```
   }
}
```

❖ ❖ ❖

9. MAKE A PRIVATE CHAT ROOM UTILIZING ARDUINO, NRF24L01 AS WELL AS PROCESSING

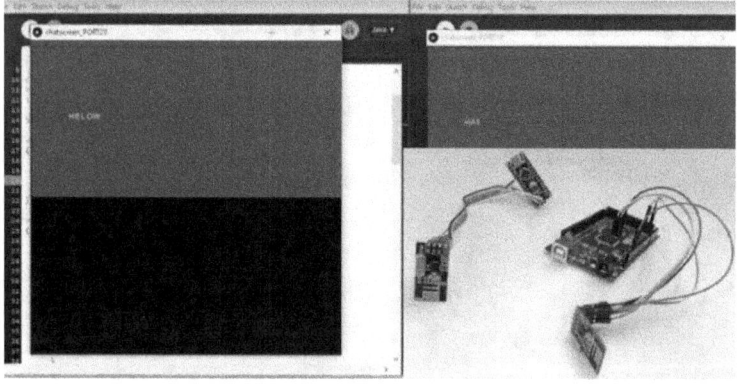

Making a Local Network to share work force and secret information's has gotten practically unthinkable for a typical man in our cutting edge world. This is primarily in light of the fact that all regular visit strategies like Whatsapp, Facebook, Hangout and nearly everything includes a web association.

Imagine a scenario where, we could share information without the vehicle of Internet.

How cool it would be on the off chance that you could speak with individuals inside your home or work place without a net pack or Internet Connection?

Consider the possibility that we could redo our talk screen with our own minds.

This is conceivable with a microcontroller and a Wireless transmission medium. This Arduino Chat Room utilizing nRF24L01 Project will control you on setting up an ease Chat Room in your neighborhood.

So we should hop in and perceive how it functions.

Working Explanation:

Fundamentally to make this thing work we will require a couple of Arduino sheets and modest remote modules. The remote modules that we will use here are nRF24L01. The purpose behind picking these modules is that these are the successors of Zigbee and is anything but difficult to work with a set up association. Additionally these modules chip away at 2.4 GHz(ISM band) with recurrence jumping spread range and stun burst alternatives which makes us feel loose of obstruction issues.

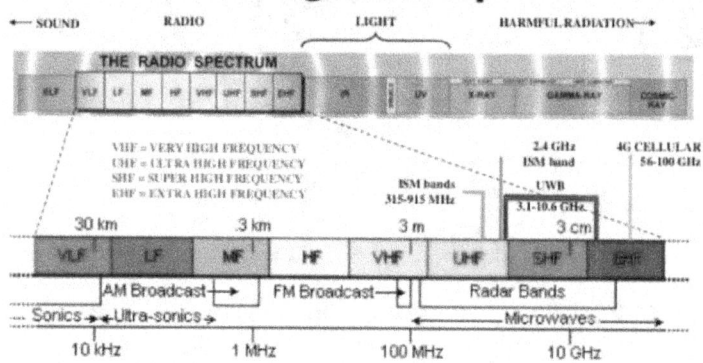

Our Arduino and NRF24L01 are associated together

to set up a Serial correspondence so they could converse with one another. The NRF24L01 are half duplex handset modules, consequently they can send and get information. The information is gathered from the client and transmitted this information can be gotten by any (or one specific) modules and show it on their screen.

But!!!!! Is it accurate to say that we are going to visit utilizing the Debug screen of Arduino? Obviously not. We are going to constructed and redo our own visit screen with assistance of 'Handling'. Handling is a product which is equipped for speaking with the Arduino utilizing UART. We will make an .exe record with Processing language, which could run on any PC with Java Runtime. So as to Chat we simply need to connect our Arduino and open this .exe record, and Booooom!! we are into our very own Privatized absolutely free visit Room.

This undertaking is restricted to simply adding two individuals to the Chat room, But the nRF24L01 has 6 Pipelines, thus there could be a limit of 6 individuals in our visit room. This burn room can work inside the 100 meter range contingent on the nRF24L01 Modules.

So how about we go shopping!!!!

Components Required:

- Arduino (any form or model) - 2Nos

- nRF24L01+ Wireless Transceiver Module - 2Nos

- 3.3 Voltage Regulator - 2Nos. (Not obligatory)

- Associating wires

- Intrigue (Lolz)

Circuit Diagram:

Arduino Mega with nRF24L01:

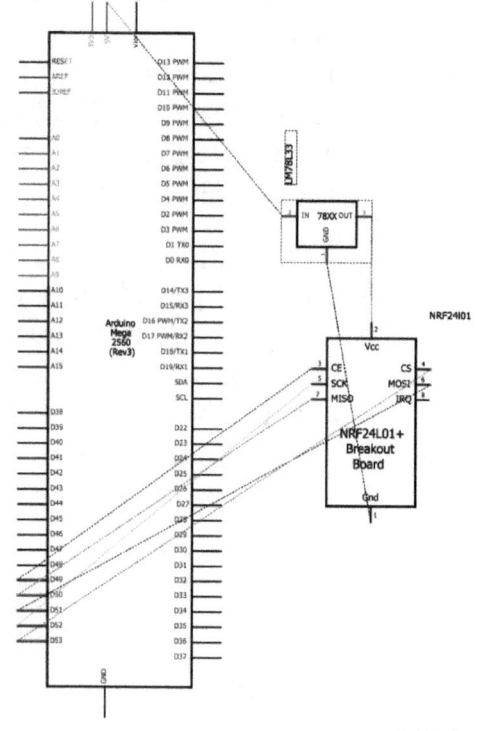

fritzing

Arduino Nano with nRF24L01:?

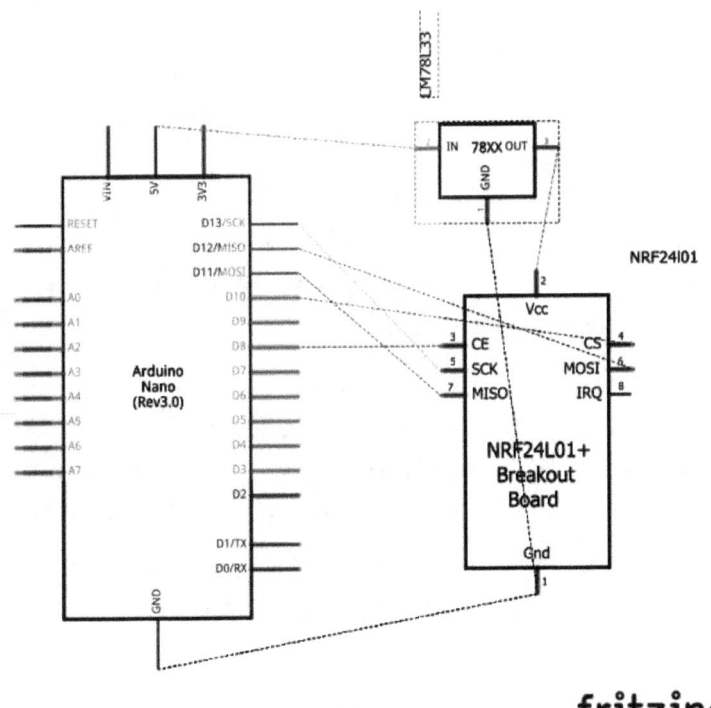

Our task doesn't include any perplexing associations. I have utilized an Arduino Mega and an Arduino Nano and their associations with nRF24L01 are appeared previously. You can utilize any Arduino Models.

Working with nRF24L01+ Wireless Transceiver Module:

Anyway so as to make our nRF24L01 to work free

from commotion we should think about the accompanying things. I have been chipping away at this nRF24L01+ for quite a while and took in the accompanying focuses can help you from getting hit on a divider.

1. The large majority of the nRF24L01+ modules in the market are phony. The modest ones that we can discover on Ebay and Amazon are the most exceedingly terrible (Don't stress, with hardly any changes we can make them work)

2. The principle issue is the power supply not your code. The vast majority of the codes online will work appropriately, I myself have a working code which I for one tried, Let me know whether you need them.

3. Focus on the grounds that the modules which are printed as NRF24L01+ are really Si24Ri (Yes a Chinese item).

4. The clone and phony modules will expend more power, subsequently don't build up your capacity circuit dependent on nRF24L01+ datasheet, in light of the fact that Si24Ri will have high current utilization about 250mA.

5. Be careful with Voltage waves and current floods, these modules are very touchy and may effectively catch fire. (;-(seared up 2 modules up until now)

6. Including a couple capacitor (10uF as well as 0.1uF) crosswise over Vcc and Gnd of the module helps in making your stock unadulterated and this works for a

large portion of the modules.

Still on the off chance that you have issues perused this.

Programming the Arduinos:

The program for both Arduino Nano and Mega will be comparative for the adjustment in CE and CS pins. I will clarify the program by parting it into little sections.

Since the Arduino and nRF24L01 conveys through the SPI we have called for SPI library. We have likewise incorporated our Maniacbug RF24 lib for our RF modules to work. Download the record from here, and add them to your IDE.

```
#include <SPI.h>

#include "RF24.h"
```

Our NRF modules are associated with stick 8 and 10 to CE and CS separately.

```
RF24 myRadio (8, 10);
```

We make an organized information bundle called bundle. The variable text[20] will be use to transmit information on air.

```
struct package

{

  char text[20]; //Text to transmit on air

};

typedef struct package Package;

Package data;
```

In the void arrangement() work, we introduce the baud rate to 9600 and arrangement our modules to 115 with MIN control utilization and 250KBPS speed. You can mess with these qualities later.

```
void setup()

{

  Serial.begin(9600);

  delay(1000);

  //Serial.print("Setup Initialized");

  myRadio.begin();
```

```
  myRadio.setChannel(115);    //115 band above
WIFI signals

  myRadio.setPALevel(RF24_PA_MIN);           //MIN
power low rage

  myRadio.setDataRate( RF24_250KBPS ) ; //Min-
imum speed

}
```

The module is made to work in transmit mode if Data is gotten through Serial cradle, else it will be in recipient mode searching for information on air. The information from client is put away in a scorch Array and sent to WriteData() to transmit them.

```
void loop()

{

  while(Serial.available()>0) //Get values from user

  {

   val = Serial.peek();

   if(index < 19) // One less than the size of the array
```

```
    {

        inChar = Serial.read(); // Read a character

        inData[index] = inChar; // Store it

        index++; // Increment where to write next

        inData[index] = '\0'; // Null terminate the
string

    }

  if(val=='#')

    {

    strcpy( data.text, inData);

    WriteData(); //Put module in Transmit mode

    while (index!=0)

    {

    inData[index] = '';

    index--;

    }
```

```
      }

  }

ReadData(); //Put module Receive mode

}
```

void WriteData() work composes the information on 0xF0F0F0F0AA address, this location is utilized as composing funnel on other module.

```
void WriteData()

{

  myRadio.stopListening(); //Stop Receiving and
start transminitng

  myRadio.openWritingPipe(0xF0F0F0F066);//
Sends data on this 40-bit address

  myRadio.write(&data, sizeof(data));

  delay(300);

}
```

void ReadData() work composes the information on

0xF0F0F0F066 this location, this location is utilized as perusing channel on other module.

```
void ReadData()

{

myRadio.openReadingPipe(1, 0xF0F0F0F0AA); //
Which pipe to read, 40 bit Address

  myRadio.startListening(); //Stop  Transminting
and start Reveicing

  if ( myRadio.available())

  {

    while (myRadio.available())

    {

      myRadio.read( &data, sizeof(data) );

    }

    Serial.println(data.text);

  }
```

```
}
```

That is it, our programming part is finished. In the event that you can't comprehend scarcely any things here, check the two projects for both the Arduinos, given in the Code area beneath, I have added remark lines to clarify things much better.

Processing Program:

'Preparing' is open source programming which is utilized by specialists for Graphics planning. This product is utilized to create programming and Android applications. It is very simple to create and particularly like the Android Development IDE. Thus I have abbreviated the clarification.

The Processing Code for both the Chat Screens is given here:

- Visit Screen 1 Processing Code

- Visit Screen 2 Processing Code

Right click on them and snap on 'Spare connect as..' to download them and open them in your PC subsequent to setting up the Arduinos. You have to introduce 'Preparing' programming to open these *.pde records and afterward 'Run' them to open the Chat Boxes. The Processing sketch for transmitter and Receiver module are indistinguishable.

In the underneath code segment the "port = new Serial(this,Serial.list()[4],9600);/Reads the fourth PORT at 9600 baudrate" is significant as it chooses from which port to information from.

```
void setup()

{

  size(510,500);

  port = new Serial(this,Serial.list()[4],9600); //
Reads the 4th PORT at 9600 baudrate

  println(Serial.list());

  background(0);

}
```

Here I have perused information from the fourth port from my Arduino.

So for Example in the event that you have COM[5] COM[2] COM[1] COM [7] COM[19]

At that point the above code will peruse information from COM[7].

Testing:

Presently since our Processing and Arduino sketch is prepared, simply transfer the program to Arduino and leave it connected to your Laptop. Open your Processing representation and start composing and press "Enter" your message will be transmitted to the next Arduino which will show the got content on another Processing application associated with other PC.

So this how you can converse with your loved ones in your neighborhood having any Internet association, utilizing this reasonable Arduino Chat Room.

Code

Code for Arduino Mega:

```
/**********
Arduino Mega with NRF24l01 for Chatroom project
**********/
#include <SPI.h>
#include "RF24.h" // Manicbug LIB to be downloaded
```

```
RF24 myRadio (49, 53); // CE to 49 and 53 to CS
struct package
{
 char text[20]; //Text to transmit on air
};
//byte addresses[][6] = {"0"};
char inData[20]; // Allocate some space for the string
char inChar; // Where to store the character read
byte index = 0; // Index into array; where to store the
character
 int val;

typedef struct package Package;
Package data;

void setup()
{
 Serial.begin(9600);
 delay(1000);
 //Serial.print("Setup Initialized");
 myRadio.begin();
  myRadio.setChannel(115);  //115 band above WIFI
signals
   myRadio.setPALevel(RF24_PA_MIN); //MIN power
low rage
   myRadio.setDataRate( RF24_250KBPS ) ;   //Min-
imum speed
}
void loop()
{
```

```
  while(Serial.available()>0) //Get values from user
{
 val = Serial.peek();
 if(index < 19) //One less than the size of the array
  {
    inChar = Serial.read(); // Read a character
    inData[index] = inChar; // Store it
    index++; // Increment where to write next
    inData[index] = '\0'; // Null terminate the string
  }
 if(val=='#')
  {
   strcpy( data.text, inData);
    WriteData(); //Put module in Transmit mode
  while (index!=0)
   {
   inData[index] = '';
   index--;
   }
  }
 }
ReadData(); //Put module Receive mode
}
void ReadData()
{
myRadio.openReadingPipe(1, 0xF0F0F0F0AA); //
Which pipe to read, 40 bit Address
 myRadio.startListening(); //Sopt Transminting and
start Reveicing
 if( myRadio.available())
```

```
{
  while (myRadio.available())
  {
   myRadio.read( &data, sizeof(data) );
  }
 // Serial.print("\nReceived:");
  Serial.println(data.text);
 }
}
void WriteData()
{
 myRadio.stopListening(); //Stop Receiving and start
transminitng
  myRadio.openWritingPipe(0xF0F0F0F066);//Sends
data on this 40-bit address
 myRadio.write(&data, sizeof(data));
  //Serial.print("\nSent:");
  //Serial.println(data.msg);
 delay(300);
}
```

Code for Arduino Nano:

```
/**********
Arduino Nano with NRF24l01 for Chatroom project
Code by B.Aswinth Raj
on 8-2016
**********/

#include <SPI.h>
#include "RF24.h" // Manicbug LIB to be downloaded
RF24 myRadio (8, 10); // CE to 8 and 10 to CS
struct package
```

```
{
 char text[20]; //Text to transmit on air
};
typedef struct package Package;
Package data;
char inData[20]; // Allocate some space for the string
char inChar; // Where to store the character read
byte index = 0; // Index into array; where to store the
character
 int val;
void setup()
{
 Serial.begin(9600);
 delay(1000);
 myRadio.begin();
  myRadio.setChannel(115);  //115 band above WIFI
signals
   myRadio.setPALevel(RF24_PA_MIN); //MIN power
low rage
 myRadio.setDataRate( RF24_250KBPS ); //Minimum
speed
 //Serial.print("Setup Initialized");
}
void loop()
{
 while(Serial.available()>0) //Get values from user
 {
  val = Serial.peek();
  if(index < 19) // One less than the size of the array
   {
```

```
    inChar = Serial.read(); // Read a character
    inData[index] = inChar; // Store it
    index++; // Increment where to write next
    inData[index] = '\0'; // Null terminate the string
  }
  if (val=='#')
  {
   strcpy( data.text, inData);
    WriteData(); //Put module in Transmit mode
  while (index!=0)
   {
   inData[index] = '';
   index--;
   }
   }
 }
ReadData(); //Put module Receive mode
}
void WriteData()
{
 myRadio.stopListening(); //Stop Receiving and start
transminitng
   myRadio.openWritingPipe( 0xF0F0F0F0AA);   //
Sends data on this 40-bit address
 myRadio.write(&data, sizeof(data));
//Serial.print("\nSent:");
//Serial.println(data.text);
 delay(300);
}
void ReadData()
```

```
{
myRadio.openReadingPipe(1,    0xF0F0F0F066);   //
Which pipe to read, 40 bit Address
 myRadio.startListening(); //Stop Transminting and
start Reveicing
 if ( myRadio.available())
 {
  while (myRadio.available())
  {
   myRadio.read( &data, sizeof(data) );
  }
  //Serial.print("\nReceived:");
  Serial.println(data.text);
 }
}
```

❖ ❖ ❖

10. ADVANCED MOBILE PHONE CONTROLLED DIGITAL CODE LOCK UTILIZING ARDUINO

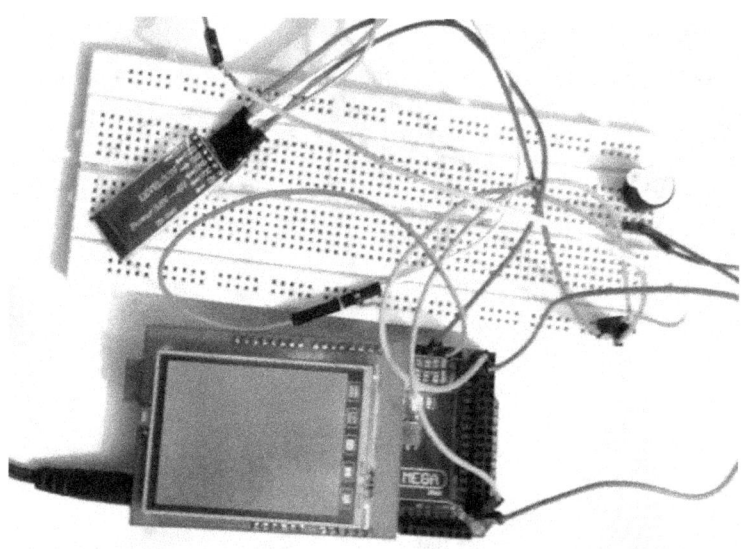

There are numerous kinds of security frameworks utilized everywhere throughout the world and Digital Code Lock is one of them. We have just secured numerous advanced locks with straightforward 16x2 LCD utilizing Arduino, Raspberry Pi, 8051 and so forth. We are gonna to fabricate a advanced cell controlled computerized Lock utilizing TFT LCD as well as Arduino Mega. This lock can be controlled remotely by means of Bluetooth, utilizing your Android telephone, inside the scope of typical Bluetooth that is 10 meters. Client needs to enter the Predefined secret key from his Android Phone, if secret key is right TFT LCD shows the "Right PASSWORD" message and in the event that secret key isn't right, at that point LCD shows "An inappropriate PASSWORD" message.

By utilizing this Lock, you can open the entryway lock, while strolling, even before coming to it. This will spare your time and you don't have to convey the keys and lock can be opened effectively with your Phone.

Required Components:

- Arduino MEGA
- USB Cable
- HC05 Bluetooth Module
- Buzzer
- Bluetooth terminal App
- Connecting wires
- Android Mobile phone
- 2.4 inch TFT LCD Shield with SPFD5408 controller
- Breadboard

Circuit Diagram and Explanation:

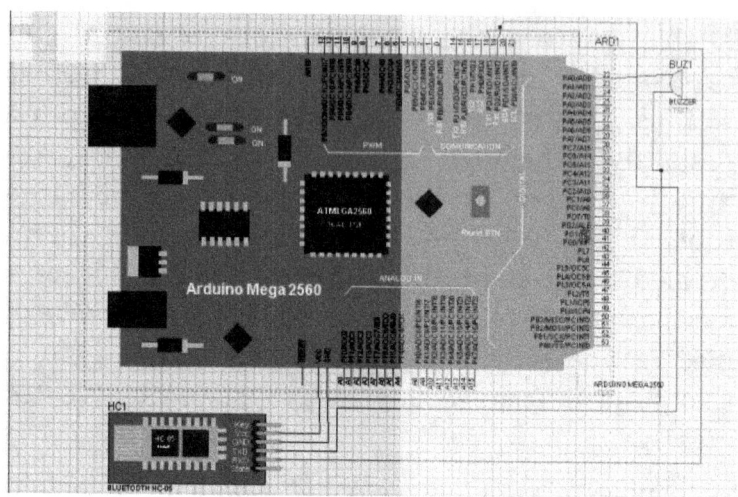

Circuit of this Smart Phone Controlled Digital Lock is basic; we just need to associate Bluetooth Module HC05 and TFT LCD Shield to the Arduino. TFT LCD shield can be effectively mounted on Arduino, we simply need coordinate the arrangement of pins and guarantee that GND and Vcc pins of Arduino ought to be mounted on GND and Vcc pins of LCD. You likewise need to introduce the Library for TFT Touch Screen LCD, get familiar with Interfacing TFT LCD with Arduino here.

HC05 is fueled by Arduino Vcc and GND Pins, TX of HC05 is associated with RX1 of Arduino and RX of HC05 is associated with TX1 of Arduino. One stick of ringer is associated with GND of Arduino and other to stick 22 of Arduino.

Configuring Bluetooth Terminal App for Arduino:

To work this advanced lock through our android PDA, first we have to introduce an android portable application named bluetooth terminal . Bluetooth Terminal App is good with Arduino. This App can be downloaded from the Google Play Store, and can be effectively arranged by following underneath Steps:

1. First download it from Google Play Store and introduce it in your Android cell phone.

2. Catalyst your 'Bluetooth controlled Digital Lock framework circuit'.

3. Open the application and go to alternative 'interface safely'.

4. You will discover HC05 gadget to combine.

5. Give 1234 passkey to interface with your Android Phone, similar to we use to associate other Bluetooth Devices.

Working Description:

In this Arduino Based Security System, we have utilized three significant parts which are Bluetooth Module HC05, Arduino Mega Board and 2.4 inch TFT LCD Shield.

Here four digit Password is entered by client through Android Smart Phone utilizing Bluetooth Terminal App and sent to the Arduino by means of Bluetooth. Arduino gets the information, sent by Android

Phone, utilizing Bluetooth Module HC05 as well as show it on SPFD5408 TFT LCD. Arduino contrasts the client entered Password and the Predefined secret word (1234), and shows the message as needs be. It shows the message "WRONG PASSWORD" if secret word don't match and show the message "Right PASSWORD" if secret phrase matches. A bell is additionally utilized for alert sign, which blares when secret key entered isn't right.

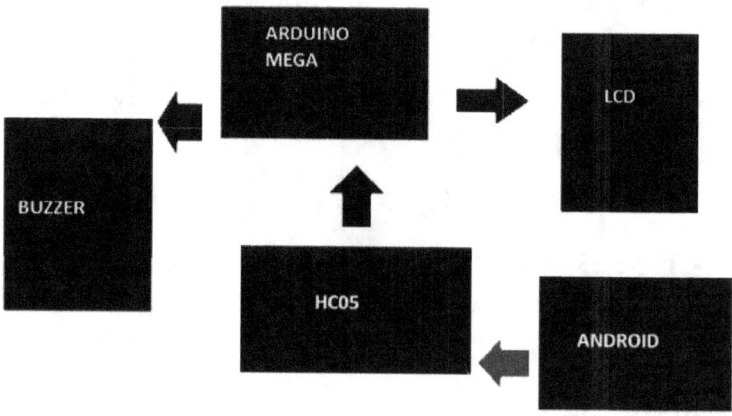

We can likewise change the secret key to our decision by changing the Arduino Code, it has been clarified in 'Programming' area beneath.

Programming Description:

To program this Bluetooth Controlled Digital Lock, we have utilized a few libraries for showing information on TFT LCD, which are given underneath. Every one of the libraries come in one rar document and can be downloaded from given this connection. Snap on 'Clone or download' and 'Download ZIP' record and add to your Arduino library envelope. This library is required for legitimate working of TFT LCD.

```
#include <SPFD5408_Adafruit_GFX.h>     // Core
graphics library

#include    <SPFD5408_Adafruit_TFTLCD.h>    //
Hardware-specific library

#include <SPFD5408_TouchScreen.h>
```

Instatement of LCD input-yield, and sequential correspondence for Bluetooth module are performed in void arrangement() circle. Stick number 22 of Arduino is interfaced to the bell and the other stick of signal is interfaced to ground of Arduino Mega. The Bluetooth module is interfaced with Serial1 port of Arduino Mega as well as controlled by 5V supply of Arduino Mega.

fillScreen() work is utilized for clearing the LCD.

```
void setup() {

// put your setup code here, to run once:
```

```
Serial.begin(9600);

Serial1.begin(9600);

tft.reset();

tft.begin(0x9341);

tft.setRotation(0);

tft.fillScreen(WHITE);

tft.setCursor(40, 50);

tft.setTextSize(2);

tft.setTextColor(BLACK);

tft.println("E N T E R* P A S S W O R D");

delay(5000);

tft.fillScreen(WHITE);

pinMode(22,OUTPUT);

}
```

In void circle() work, setTextSize(4) sets the size of content and setTextColor(colorName) sets the shade of content. The arr[] is the exhibit where we have put away the predefined four digit secret key and the Input[] is the cluster wherein we have put away the secret word entered by client from Android Phone. In the event that secret key entered is same as the secret word put away, at that point the LCD will show "Right PASSWORD" message and If it isn't the equivalent for example 1234, at that point the LCD will show "WRONG PASSWORD" message and the stick associated with signal turns out to be high and the ringer blares.

```
void loop() {

tft.setTextSize (4);

tft.setTextColor(CYAN);

// put your main code here, to run repeatedly:

if (Serial1.available() > 0)

{

flag = 1;

char c = Serial1.read();

if (flag == 1)
```

```
{

    input[i] = c;

    ..... .....

    ..... ......
```

We can further, adjust the arr[] cluster to change our preferred secret phrase rather than '1234'. We can likewise change the no. of characters in the secret phrase by changing the length of arr[] and input[] clusters.

```
char arr[4] = {'1', '2', '3', '4'};

char input[4];
```

As indicated by the changed length of secret key we have to change the if condition in void circle() work.

```
if (arr[0] == input[0] && arr[1] == input[1] && arr[2]
== input[2] && arr[3] == input[3])
```

Further we can interface an Electronic Door Lock (effectively accessible on the web) in this task. It have an Electro magnet which keeps the Door bolted when there is no current gone through the Lock (open circuit), and when some current went through it, the lock gets opened and entryway can be opened. We simply need to change the Code in like manner, and we are prepared to Open entryway Lock with our Phone. Check this mutual venture audit: Arduino RFID Door Lock to get progressively about Electronic Door Lock.

Code

```
#include <SPFD5408_Adafruit_GFX.h>     // Core graphics library
#include <SPFD5408_Adafruit_TFTLCD.h> // Hardware-specific library
#include <SPFD5408_TouchScreen.h>
//*** SPFD5408 change -- End
#if defined(__SAM3X8E__)
#undef __FlashStringHelper::F(string_literal)
#define F(string_literal) string_literal
#endif
#define YP A1 // must be an analog pin, use "An" notation!
#define XM A2 // must be an analog pin, use "An" notation!
#define YM 7  // can be a digital pin
#define XP 6  // can be a digital pin
// Original values
//#define TS_MINX 150
//#define TS_MINY 120
//#define TS_MAXX 920
//#define TS_MAXY 940
// Calibrate values
```

```
#define TS_MINX 125
#define TS_MINY 85
#define TS_MAXX 965
#define TS_MAXY 905
```

// For better pressure precision, we need to know the resistance
// between X+ and X- Use any multimeter to read it
// For the one we're using, its 300 ohms across the X plate

```
TouchScreen ts = TouchScreen(XP, YP, XM, YM, 300);
#define LCD_CS A3
#define LCD_CD A2
#define LCD_WR A1
#define LCD_RD A0
// optional
#define LCD_RESET A4
```

// Assign human-readable names to some common 16-bit color values:

```
#define BLACK   0x0000
#define BLUE    0x001F
#define RED     0xF800
#define GREEN   0x07E0
#define CYAN    0x07FF
#define MAGENTA 0xF81F
#define YELLOW  0xFFE0
#define WHITE   0xFFFF
char arr[4] = {'1', '2', '3', '4'};
char input[4];
char  messageflagarr[4]; char result = '0', flag = 0, passwordflag = 0;
```

```
int x = 40, i = 0;
Adafruit_TFTLCD tft(LCD_CS, LCD_CD, LCD_WR,
LCD_RD, LCD_RESET);
void setup() {
// put your setup code here, to run once:
Serial.begin(9600);
Serial1.begin(9600);
tft.reset();
tft.begin(0x9341);
tft.setRotation(0);
tft.fillScreen(WHITE);
tft.setCursor(40, 50);
tft.setTextSize(2);
tft.setTextColor(BLACK);
tft.println("E N T E R* P A S S W O R D");
delay(5000);
tft.fillScreen(WHITE);
pinMode(22,OUTPUT);
}
void loop() {
tft.setTextSize(4);
tft.setTextColor(CYAN);
// put your main code here, to run repeatedly:
if(Serial1.available() > 0)
{
flag = 1;
char c = Serial1.read();
if(flag == 1)
{
input[i] = c;
```

```
 i++;
 tft.println(c);
 delay(2000);
 if(i == 4)
 {
  passwordflag = 1;
 }
 flag = 0;
 }
 if(passwordflag == 1)
 {
  if (arr[0] == input[0] && arr[1] == input[1] && arr[2]
== input[2] && arr[3] == input[3])
  {
  tft.setCursor (40, 50);
  tft.setTextSize (2);
  tft.setTextColor(BLACK);
  tft.println("CORRECT PASSWORD");
  delay(6000);
  tft.fillScreen(WHITE);
  i=0;
  }
  else
  {
  tft.setCursor (40, 50);
  tft.setTextSize (2);
  tft.setTextColor(BLACK);
  tft.println("WRONG PASSWORD");
  digitalWrite(22,HIGH);
  delay(500);
```

```
digitalWrite(22,LOW);
delay(500);
digitalWrite(22,HIGH);
delay(500);
digitalWrite(22,LOW);
delay(500);
digitalWrite(22,HIGH);
delay(500);
digitalWrite(22,LOW);
delay(500);
digitalWrite(22,HIGH);
delay(500);
digitalWrite(22,LOW);
tft.fillScreen(WHITE);
digitalWrite(22,LOW);
i=0;
}
passwordflag=0;
}
}
}
```

THANK YOU !!!